10 Florida FAST Grade 7 Math Practice Tests

The Ultimate Test Prep Collection with Answer Explanations

Dr. A. Nazari

Copyright © 2026 Dr. A. Nazari

Published by View Math Education

ViewMath.com

All rights reserved. No part of this publication may be reproduced, distributed, or transmitted in any form or by any means, including photocopying, recording, or other electronic or mechanical methods, without the prior written permission of the author, except in the case of brief quotations embodied in critical reviews and certain other noncommercial uses permitted by copyright law, including Section 107 or 108 of the 1976 United States Copyright Act.

The information in this book is distributed on an "as is" basis, without warranty. While every precaution has been taken in the preparation of this work, neither the author nor the publisher shall have any liability to any person or entity with respect to any loss or damage caused or alleged to be caused directly or indirectly by the information contained in this book.

Copyright © 2026

10 Practice Tests

Grade 7 Math — Engineered for Mastery

Welcome to **The Architect's Workshop.**

Ten practice tests. Ten floors of a tower you're about to build. Each test lays another level of understanding — from foundation to capstone.

- **Foundation (Tests 1–3):** Learn the blueprints
- **Framework (Tests 4–6):** Build structural strength
- **Finishing (Tests 7–9):** Refine under pressure
- **Capstone (Test 10):** The final inspection

Precision. Practice. Perfection.

> **" Every great structure is built one floor at a time. Ten levels of practice makes your math rock-solid. "**

© VisualMath.com — All Rights Reserved

The Architectural Plan

A 4-phase construction plan for complete mastery

Phase I: Foundation (Tests 1–3)

Untimed. Explore the test format and question types. Read every answer explanation after each test. Goal: understand the blueprint before you build.

Phase II: Framework (Tests 4–6)

Add a timer (70 minutes). Focus on the topics that gave you trouble in Phase I. Practice showing complete work and labeling units. Goal: build structural strength.

Phase III: Finishing (Tests 7–9)

Full timed conditions (60 minutes). Simulate the real exam. Review only the questions you missed — don't re-study what you already know. Goal: refine under real pressure.

Phase IV: Capstone (Test 10)

Your final inspection. Full exam conditions. This is the capstone — compare with Test 1 and measure your entire growth arc. Goal: prove your mastery.

Blueprint Specifications

Multiple Choice — select the single best answer

Multi-Select — choose ALL that apply

Short Answer — show your process

Open Response — explain and justify

⎾ Engineering Principles ⏌

The D.R.A.F.T. Method

D **Define** What exactly is the question asking? Write it in your own words.

R **Retrieve** Pull out given information. List numbers, units, and relationships.

A **Assemble** Choose the right formula or strategy. Set up the equation or proportion.

F **Figure** Compute step by step. Show every operation. Label units.

T **Test** Does the answer make sense? Re-read the question. Verify the units.

Seven Precision Rules

1. **Read twice, solve once.** The first read gives context; the second reveals what to compute.

2.

3. **Estimate before calculating.** A quick mental approximation catches major errors.

4.

5. **Track your signs.** Rational number operations are the #1 error source in Grade 7.

6.

7. **Never submit blanks.** On open response, even a partial setup can earn credit.

Timing Blueprint

Tests 1–3: **Untimed** (learn the blueprints) > Tests 4–6: **70 min** (build speed) > Tests 7–10: **60 min** (exam conditions)

Every great structure starts with a solid plan. Study the blueprints, learn from each test, and build something extraordinary.

Get Online

Find more at
ViewMath.com/FL-Grade7

🧰 *The Drafting Toolkit* 🧰

Prepare your station before every construction session

Required Equipment

	Precision Pencils	Two or more #2 pencils, kept sharp. A dull tool produces dull work.
	Quality Eraser	Clean, smudge-free corrections. Architects revise — so will you.
	Scratch Paper	Your drafting workspace. All calculations happen here first.
	Ruler / Protractor	Precision tools for scale drawings, angle measures, and geometry.
	Timer	Begin using from Phase II (Test 4) onward.
	Quiet Workspace	A clean, well-lit station free from distractions.

Approved Materials

- ✔ Pencil and eraser
- ✔ Scratch paper (provided on test day)
- ✔ Ruler or protractor (if specified)
- ✔ Reference sheet (if provided)

Restricted Materials

- ✘ Calculator (unless specified)
- ✘ Electronic devices
- ✘ Notes or reference materials
- ✘ Communication with others

♥ Project Manager's Guide (for Parents & Guardians)

Ten practice tests provide the most comprehensive preparation available. The 4-phase structure (Foundation → Framework → Finishing → Capstone) ensures skills build progressively and durably.

Keys to a successful build:

- *Space tests 2–3 days apart — never more than one per day*
- *Phases I–II: review answers together and discuss strategies*
- *Phases III–IV: let them work independently, then debrief results*
- *Compare Test 1 with Test 10 to celebrate the full construction arc*

Find more at
ViewMath.com/FL-Grade7

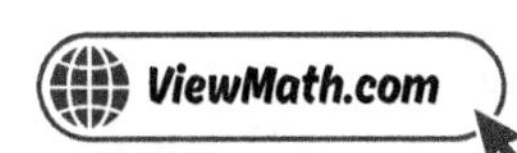

Symbol	Name	What It Means	
$(\ )$	Parentheses	Do this part first.	$(3+4) \times 2 = 14$
10^3	Exponent	Multiply the base by itself that many times. $10^3 = 1,000$	
$\dfrac{a}{b}$	Fraction	a parts out of b equal parts; also means $a \div b$.	
$\dfrac{7}{3}$	Improper Fraction	Numerator $\geq$ denominator.	$\frac{7}{3} = 2\frac{1}{3}$
0.45	Decimal	A number with a decimal point.	$0.45 = \frac{45}{100}$
$> < =$	Comparison	Greater than, less than, equal to.	$0.5 > 0.35$
$(3,5)$	Ordered Pair	A point on the coordinate plane: (x, y).	

🖩 Key Formulas

- **Volume of a rectangular prism:**

  $V = l \times w \times h$

- **Order of operations:**

  Parentheses → Exponents → Multiply/Divide →

  Add/Subtract

- **Powers of 10:**

  $10^1 = 10 \quad 10^2 = 100$

  $10^3 = 1,000 \quad 10^4 = 10,000$

- **Fraction as division:**

  $\frac{a}{b} = a \div b$

▦ Place Value Chart

Millions	1,000,000	**Decimals**	
Hundred-Thousands	100,000	Tenths	0.1
Ten-Thousands	10,000	Hundredths	0.01
Thousands	1,000	Thousandths	0.001
Hundreds	100		
Tens	10		
Ones	1		

Each place is $10\times$ the place to its right, and $\frac{1}{10}$ of the place to its left.

📖 Key Math Vocabulary

- **Sum** — the result of addition
- **Difference** — the result of subtraction
- **Product** — the result of multiplication
- **Quotient** — the result of division
- **Remainder** — what's left over after dividing
- **Factor** — a number you multiply
- **Expression** — numbers and operations without $=$
- **Equation** — a math sentence with $=$
- **Numerator** — the top number of a fraction
- **Denominator** — the bottom number of a fraction
- **Mixed number** — a whole number + a fraction
- **Equivalent fractions** — fractions with equal value
- **Decimal** — a number written with a decimal point
- **Volume** — the space inside a 3D shape
- **Coordinate plane** — a grid with x and y axes
- **Ordered pair** — (x, y) location on the plane

🔍 Word Problem Clue Words

- **Add** $(+)$: total, altogether, combined, sum, increase, more than
- **Subtract** $(-)$: difference, how many more, fewer, remain, decrease, left
- **Multiply** $(\times)$: each, every, groups of, times, product, per, of (with fractions)
- **Divide** $(\div)$: share equally, split, each group, how many groups, quotient, per

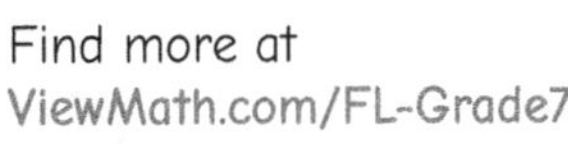

Find more at
ViewMath.com/FL-Grade7

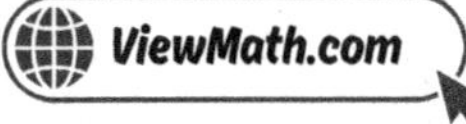

⬛ Multiplication Table ⬛

You may use this table during your practice tests!

×	1	2	3	4	5	6	7	8	9	10	11
1	1	2	3	4	5	6	7	8	9	10	11
2	2	4	6	8	10	12	14	16	18	20	22
3	3	6	9	12	15	18	21	24	27	30	33
4	4	8	12	16	20	24	28	32	36	40	44
5	5	10	15	20	25	30	35	40	45	50	55
6	6	12	18	24	30	36	42	48	54	60	66
7	7	14	21	28	35	42	49	56	63	70	77
8	8	16	24	32	40	48	56	64	72	80	88
9	9	18	27	36	45	54	63	72	81	90	99
10	10	20	30	40	50	60	70	80	90	100	110
11	11	22	33	44	55	66	77	88	99	110	121

💡 How to Use This Table

To find **4 × 7**:

1. Find **4** in the left column (blue).
2. Find **7** in the top row (blue).
3. Follow the row and column until they meet: the answer is **28**!

ⓘ Tip: You can also use this table for division! If you know $28 \div 4 =$?, find 28 in the 4's row. The column header gives you the answer: **7**!

🏢 Construction Log 🏢

Record each floor and watch your tower rise

Architect: _______________________________ **Project Start:** _________________

📐 **Floor 1**   Date: __________   Score: _____ / _____   %: _____   ★★★★★
📐 **Floor 2**   Date: __________   Score: _____ / _____   %: _____   ★★★★★
📐 **Floor 3**   Date: __________   Score: _____ / _____   %: _____   ★★★★★

📐 **Floor 4**   Date: __________   Score: _____ / _____   %: _____   ★★★★★
📐 **Floor 5**   Date: __________   Score: _____ / _____   %: _____   ★★★★★
📐 **Floor 6**   Date: __________   Score: _____ / _____   %: _____   ★★★★★

📐 **Floor 7**   Date: __________   Score: _____ / _____   %: _____   ★★★★★
📐 **Floor 8**   Date: __________   Score: _____ / _____   %: _____   ★★★★★
📐 **Floor 9**   Date: __________   Score: _____ / _____   %: _____   ★★★★★

📐 **Floor 10**   Date: __________   Score: _____ / _____   %: _____   ★★★★★

Certification Level: Apprentice (0–39%) Technician (40–59%) Engineer (60–79%) **Master Architect** (80–100%)

Record your certification level after each test!

🔧 Under Construction

✅ Construction Complete

Find more at
ViewMath.com/FL-Grade7

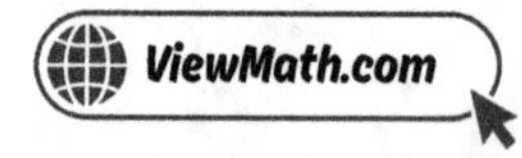

YOUR ONLINE COMPANION

Continue Learning at ViewMath Academy!

For Parents, Teachers & Students

Great job on the practice tests! Want to keep improving? ViewMath Academy is your **free online companion** to this book.

- **Score Analyzer** — Enter your answers and instantly see which topics need more practice
- **Interactive Lessons** — Review the concepts behind each question with clear explanations
- **Adaptive Quizzes** — Practice your weak topics with questions that match your level
- **Progress Tracking** — See your mastery grow across all Grade 7 math topics
- **Personalized Dashboard** — A learning plan tailored just for you

Scan to visit ViewMath Academy

ViewMath.com/FL-Grade7

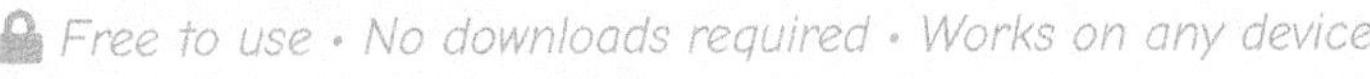 Free to use · No downloads required · Works on any device 

 Let's learn and have fun!

Practice Test 1

 30 Questions

✏️ Before You Start ✏️

- ✔ **Read each question carefully** before choosing your answer.
- ✔ **Show your work** on scratch paper when you need to.
- ✔ **Skip hard questions** and come back to them later.
- ✔ **Check your answers** when you're done.
- ✔ **Take your time** — there's no rush!

 You've Got This! 

Do your best and show what you know!

1. The table shows a proportional relationship. Find k and the missing value.

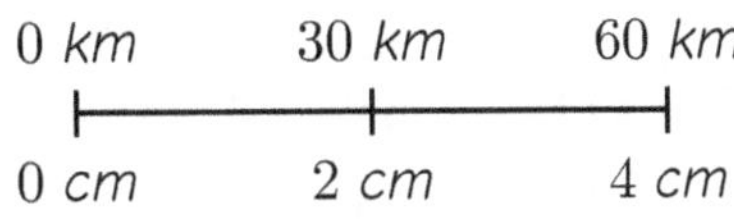

x	y
3	7.5
?	15
10	25

Your Answer

2. A map uses the scale shown below. If two cities are 6.5 cm apart on the map, what is the actual distance?

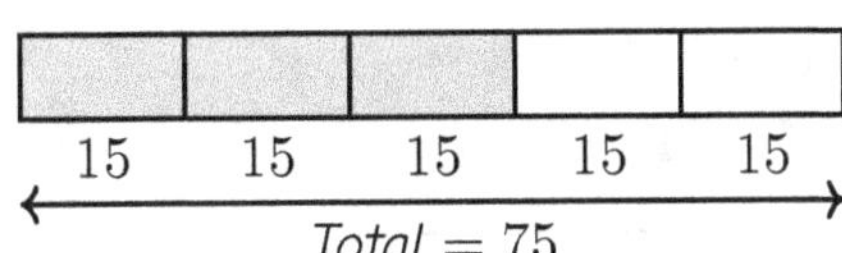

(A) 90 km

(B) 95 km

(C) 97.5 km

(D) 100 km

3. The tape diagram below represents a whole quantity. What percent of the total is the shaded portion?

(A) 40%

(B) 45%

(C) 50%

(D) 60%

4. A school has 500 students. If 68% eat in the cafeteria, how many eat there? Use a proportion.

(A) 320

(B) 340

(C) 350

(D) 368

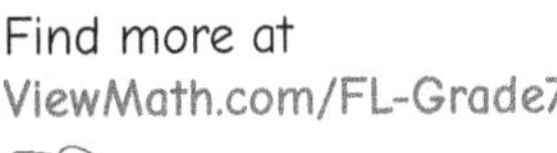

Find more at
ViewMath.com/FL-Grade7

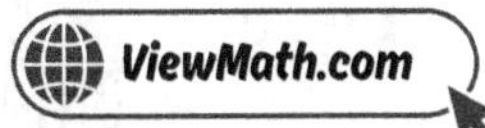

5. The bar graph shows the interest earned on three different deposits over 2 years. Which deposit had the highest interest rate?

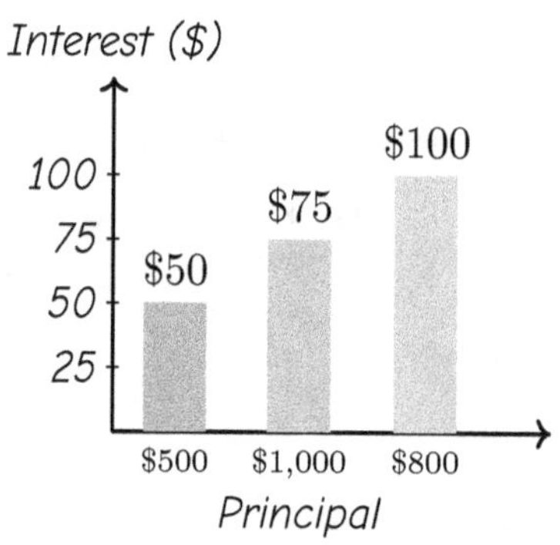

(A) Deposit of $500 (earned $50)

(B) Deposit of $1,000 (earned $75)

(C) Deposit of $800 (earned $100)

(D) All have the same rate

6. A jogger estimated running 5 miles but actually ran 4.8 miles. What is the percent error (rounded to the nearest tenth)?

Your Answer:

7. Which number line correctly models $(-4) + (-3)$?

A:
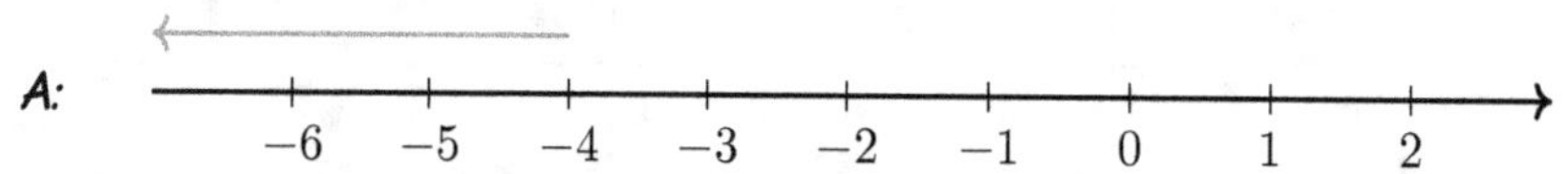

B:
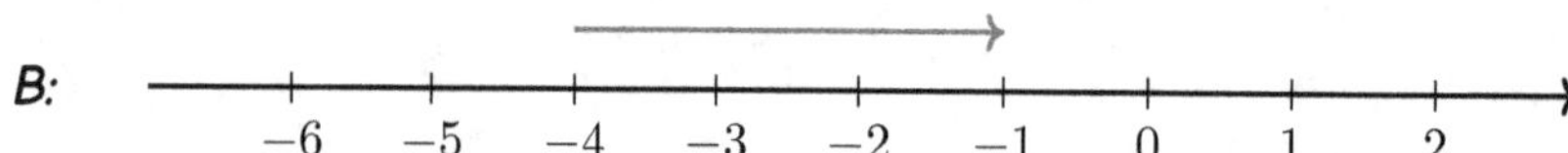

(A) Model A

(B) Model B

(C) Both are correct

(D) Neither is correct

Find more at
ViewMath.com/FL-Grade7

ViewMath.com

8. *What is $0 - (-6)$?*

(A) -6 (B) 6

(C) 0 (D) -12

9. *Factor $-8x + 12y - 4$*

Your Answer:

10. *Simplify $(-3x + 5) + (8x - 12)$.*

Your Answer:

11. *Look at the number line. The expression $3(x + 2)$ equals the distance from A to B. What is x?*

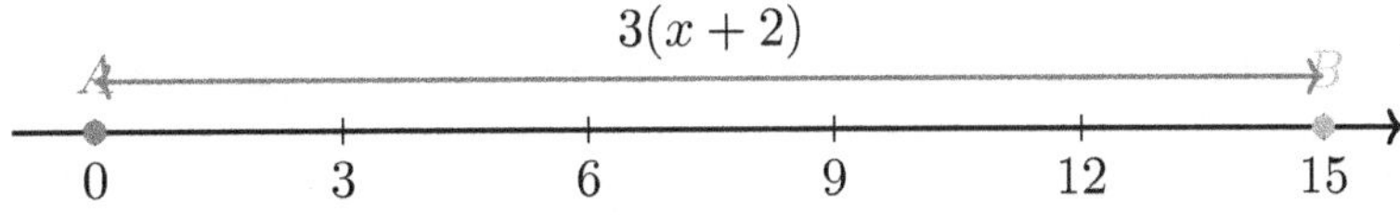

(A) $x = 5$ (B) $x = 7$

(C) $x = 3$ (D) $x = 13$

12. *A number line shows a point at x. The equation $\dfrac{x}{4} + 1.5 = 5$ tells you how to find x. Solve and identify where x is on the number line.*

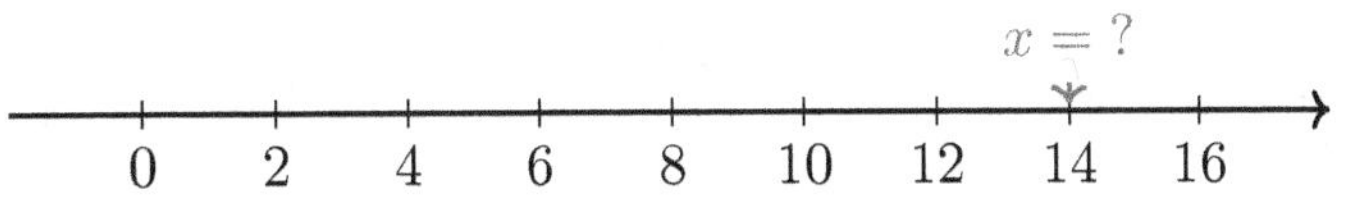

Your Answer:

Find more at
ViewMath.com/FL-Grade7

ViewMath.com

13. *You earn $9 per hour and need more than $100 for a purchase. You already have $28. Write and solve an inequality for the number of hours h you need to work.*

Your Answer:

14. *What type of circle is used to graph $x > 3$ on a number line?*

(A) *Closed circle at 3*

(B) *Open circle at 3*

(C) *Closed circle at 0*

(D) *Open circle at 0*

15. *A drawing uses 1 cm = 7 m. You redraw at 1 cm = 7 m (the same scale). A line is 5 cm long. How long is it on the new drawing?*

Your Answer:

16. *Two sides of a triangle are 9 cm and 12 cm with an included angle of 60°. What type of condition is this?*

(A) *SSS*

(B) *SAS*

(C) *ASA*

(D) *AAA*

17. *Name two different cross-section shapes you can get from slicing a cube.*

Your Answer:

18. *Two vertical angles are $(7x + 1)°$ and $(5x + 13)°$. Find x and the angle measure.*

Your Answer:

Find more at
ViewMath.com/FL-Grade7

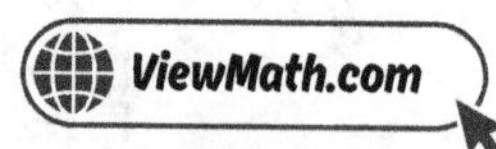

19. A circular garden has a radius of 8 feet. A path goes straight through the center from one side to the other. How long is the path?

(A) 4 feet

(B) 8 feet

(C) 16 feet

(D) 24 feet

20. A rectangular prism has a surface area of 94 cm². Its length is 5 cm and width is 3 cm. What is its height?

(A) 2 cm

(B) 3 cm

(C) 4 cm

(D) 5 cm

21. A cube has a volume of 64 cm³. What is the side length?

(A) 2 cm

(B) 4 cm

(C) 8 cm

(D) 16 cm

22. A cylinder has a volume of 314 cm³ and a radius of 5 cm. What is the height? Use $\pi \approx 3.14$.

(A) 2 cm

(B) 4 cm

(C) 6 cm

(D) 10 cm

23. A company wants to know if customers like a new product. They survey only customers who bought the product last week. What is the problem with this sample?

(A) The sample is too large

(B) The sample is biased toward recent buyers

(C) The sample is random

(D) There is no problem

Find more at
ViewMath.com/FL-Grade7

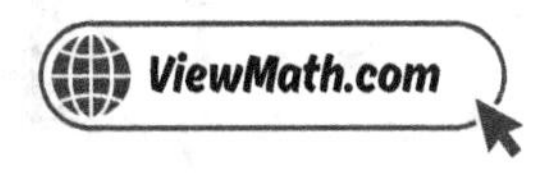

24. *Group X: mean 45, range 12. Group Y: mean 52, range 12. What can you say about these groups?*

(A) *They have the same center and different spread*

(B) *They have different centers and the same spread*

(C) *They are identical*

(D) *Group X is more spread out*

25. *The MAD for data set $\{4, 6, 8, 10, 12\}$ with mean 8 is:*

(A) 2

(B) 2.4

(C) 3

(D) 8

26. *Use the stem-and-leaf plot to answer the question.*

Stem	Leaf
1	5 8
2	0 3 3 7
3	1 5 9
4	2

Key: 1 | 5 means 15

How many data values are greater than 25?

(A) 3

(B) 4

(C) 5

(D) 6

27. *A class of 40 students: 16 prefer dogs, 12 prefer cats, 8 prefer fish, 4 prefer birds. What percent prefer dogs?*

(A) 16%

(B) 25%

(C) 40%

(D) 80%

Get Online

Find more at
ViewMath.com/FL-Grade7

ViewMath.com

28. A probability model has outcomes X, Y, and Z. $P(X) = 0.15$ and $P(Y) = 0.45$. What is $P(Z)$?

Your Answer:

29. A spinner has 4 equal sections $(1, 2, 3, 4)$ and a die is rolled. What is the probability that both show a 3?

(A) $\frac{1}{10}$ (B) $\frac{1}{24}$

(C) $\frac{2}{10}$ (D) $\frac{1}{4}$

30. A spinner has a 60% chance of landing on blue. In a simulation of 80 spins, how many times would you expect blue to appear?

Your Answer:

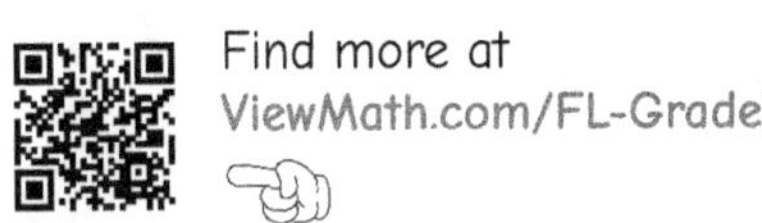
Find more at
ViewMath.com/FL-Grade7

ViewMath.com

 # End of Practice Test 1 

Great job finishing the test!

 My Score

I got _____________ out of 30 questions right.

*Check your answers in the **Answer Key** at the back of the book.*

💡 *Review any questions you missed. That's how we learn!*

📊 Check Your Score Online!

Visit **ViewMath Academy** to enter your answers and see which topics you need to review. You can also explore lessons, take quizzes, track your scores, and save your progress!

viewmath.com/score/7.1.FL.16

Or go to viewmath.com/score and enter code: 7.1.FL.16

2

Practice Test 2

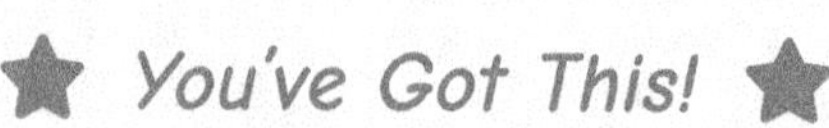

30 Questions

✏️ Before You Start ✏️

- ✓ **Read each question carefully** before choosing your answer.
- ✓ **Show your work** on scratch paper when you need to.
- ✓ **Skip hard questions** and come back to them later.
- ✓ **Check your answers** when you're done.
- ✓ **Take your time** — there's no rush!

⭐ You've Got This! ⭐

Do your best and show what you know!

1. The table below shows a proportional relationship between gallons of gas and miles driven. What is k and what does it mean?

Gallons (x)	Miles (y)
2	54
5	135
8	216

(A) $k = 2$; uses 2 gallons per trip

(B) $k = 27$; drives 27 miles per gallon

(C) $k = 54$; drives 54 miles on 2 gallons

(D) $k = 135$; drives 135 miles on 5 gallons

2. The double number line below relates inches on a blueprint to actual feet. A hallway measures 3.5 inches on the blueprint. What is the actual length?

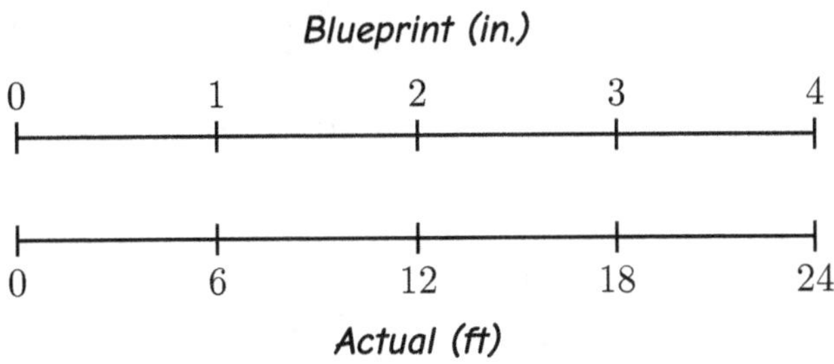

Your Answer:

3. The circle graph shows how 200 students get to school. How many students take the bus?

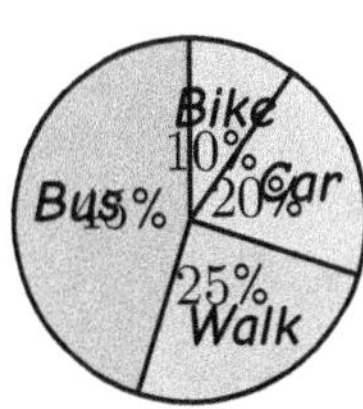

Your Answer:

4. Which proportion shows that 42 is 60% of some number w?

(A) $\dfrac{42}{w} = \dfrac{60}{100}$

(B) $\dfrac{w}{42} = \dfrac{60}{100}$

(C) $\dfrac{60}{42} = \dfrac{w}{100}$

(D) $\dfrac{42}{60} = \dfrac{w}{100}$

5. What is the simple interest on \$900 at 8% for 6 months?

(A) \$36

(B) \$72

(C) \$108

(D) \$432

6. Two students estimated the number of beans in a jar. The actual count was 200. Student A guessed 180; Student B guessed 220. Who had the greater percent error?

(A) Student A

(B) Student B

(C) Same percent error

(D) Cannot be determined

7. What is $(-12) + 7$?

(A) -19

(B) 19

(C) -5

(D) 5

8. What is $(-6) - (-6) - 6$?

Your Answer:

9. Factor $-9x - 6$.

(A) $-3(3x - 2)$

(B) $-3(3x + 2)$

(C) $3(-3x - 2)$

(D) $-9(x + 6)$

Find more at
ViewMath.com/FL-Grade7

ViewMath.com

10. *Simplify* $(4a + 3b) - (2a - 5b)$.

Your Answer:

11. Three friends split the cost of a gift equally. Each person also pays $4 for wrapping. Each person pays $16 total. What is the cost c of the gift?

(A) $c = 48$

(B) $c = 36$

(C) $c = 12$

(D) $c = 60$

12. *Solve* $-0.3x + 4.2 = 1.8$.

Your Answer:

13. The table shows ticket prices for groups. A school has $200 to spend and must also pay a $35 bus fee. Write and solve an inequality to find the maximum number of students s who can go.

Group Size	Price per Student
1–10	$12
11–25	$9
26+	$7

Assume the group qualifies for the $9 rate.

Your Answer:

14. What inequality is shown by a closed circle at -6 with shading to the right?

Your Answer:

Find more at
ViewMath.com/FL-Grade7

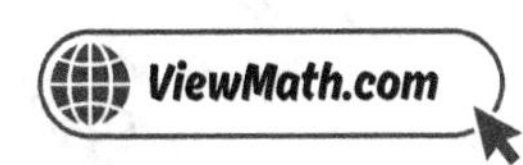

15. When you redraw a scale drawing at a smaller scale (more real units per cm), the new drawing is:

(A) Larger than the original drawing

(B) Smaller than the original drawing

(C) The same size as the original drawing

(D) A different shape than the original drawing

16. Two angles of a triangle are 35° and 75°. What is the third angle?

(A) 60°

(B) 70°

(C) 80°

(D) 110°

17. You slice a cube with a vertical cut from one face to the opposite face, perpendicular to the base. What is the most likely cross-section shape?

(A) Square or rectangle

(B) Circle

(C) Triangle

(D) Hexagon

18. Two angles are supplementary. One is 25° more than the other. What are the two angles?

(A) 65° and 115°

(B) 77.5° and 102.5°

(C) 80° and 100°

(D) 75° and 105°

19. A circle has a diameter of 5.6 inches. What is the radius?

Your Answer:

20. How many faces does a rectangular prism have?

(A) 4

(B) 5

(C) 6

(D) 8

Find more at
ViewMath.com/FL-Grade7

21. A moving box is 2 ft long, 2 ft wide, and 3 ft tall. How many of these boxes can fit in a space that is 6 ft by 4 ft by 3 ft?

(A) 3

(B) 6

(C) 12

(D) 18

22. Find the surface area of a cylinder with radius 7 cm and height 10 cm. Use $\pi \approx \frac{22}{7}$.

Your Answer:

23. A random sample means that:

(A) Only the best members are chosen

(B) The largest possible group is chosen

(C) Every member of the population has an equal chance of being selected

(D) Only volunteers are selected

24. Two groups have means of 50 and 54, and both have MAD = 8. Express the difference in means as a multiple of MAD. Is the difference meaningful?

Your Answer:

Find more at
ViewMath.com/FL-Grade7

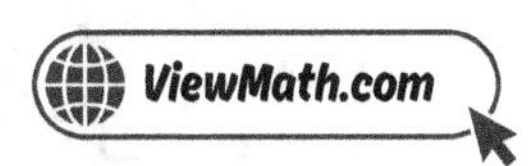

25. *The box plots show quiz scores (out of 20) for two classes.*

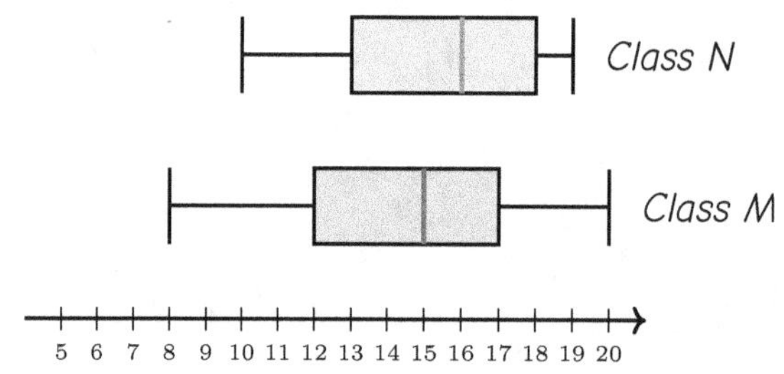

What is the IQR for each class?

(A) *Class M: 5, Class N: 5*

(B) *Class M: 12, Class N: 9*

(C) *Class M: 5, Class N: 9*

(D) *Class M: 12, Class N: 5*

26. *The stem-and-leaf plot shows the ages of people at a family reunion.*

Stem	Leaf
0	5 8 9
1	2 4 6
2	0 5 8
3	2 5 7 9
4	1 5
5	0 8
6	3 7
7	2

Key: 3 | 2 *means 32 years old*

Find the range and median of the ages.

Your Answer:

Find more at
ViewMath.com/FL-Grade7

ViewMath.com

27. A circle graph represents:

(A) How data changes over time

(B) How a whole is divided into parts

(C) The range of a data set

(D) Individual data points

28. A student creates a probability model for a spinner. $P(red) = 0.25$, $P(blue) = 0.25$, $P(green) = 0.25$, $P(yellow) = 0.25$. After 200 spins, the results are: red $= 55$, blue $= 48$, green $= 52$, yellow $= 45$. Is the model a good fit?

(A) No, the model is completely wrong

(B) Yes, the observed frequencies are reasonably close to 50 each

(C) No, because the results are not exactly 50 each

(D) Yes, because 200 is divisible by 4

29. The tree diagram shows the outcomes for flipping two coins. What is the probability of getting at least one tail?

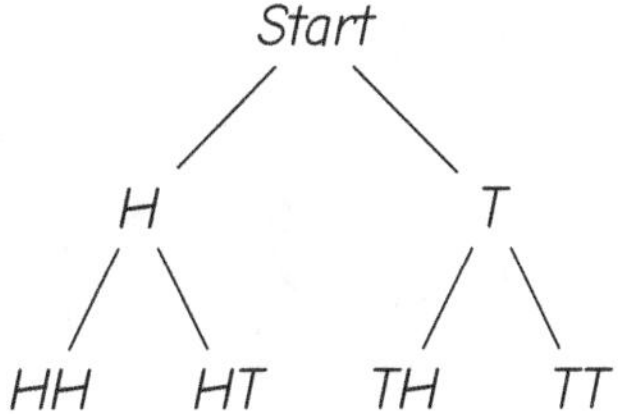

(A) $\frac{1}{4}$

(B) $\frac{1}{2}$

(C) $\frac{3}{4}$

(D) $\frac{2}{4}$

Find more at
ViewMath.com/FL-Grade7

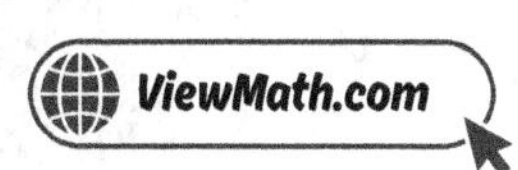

30. *The bar graph shows the results of simulating a spinner 100 times. The spinner is supposed to land on each color with equal probability. Based on the simulation, does the spinner appear fair?*

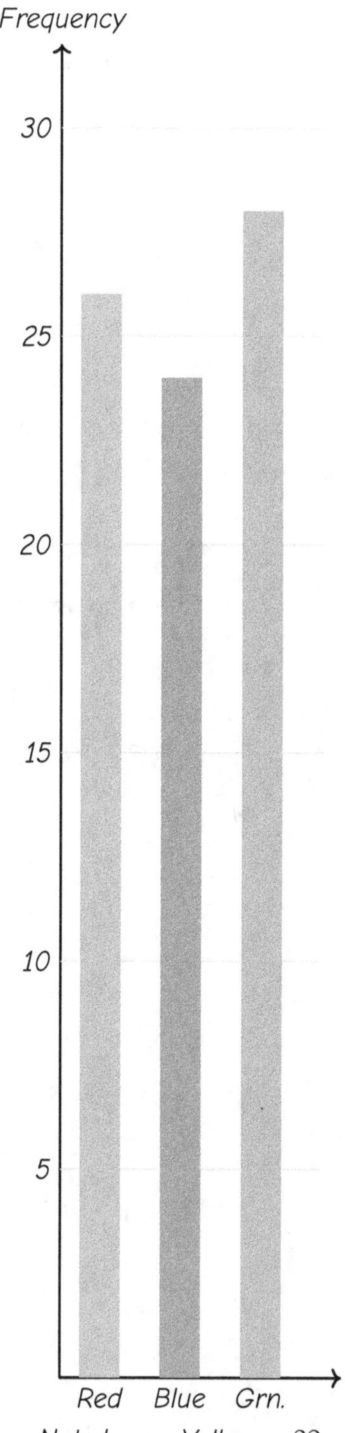

(A) *No, because the bars are different heights*

(B) *Yes, the frequencies* $(26, 24, 28, 22)$ *are all roughly close to* 25

Green is clearly the most likely color

(D) *Cannot be determined from the graph*

 *End of Practice Test 2* 

Great job finishing the test!

 **My Score**

I got _____________ out of 30 questions right.

*Check your answers in the **Answer Key** at the back of the book.*

Review any questions you missed. That's how we learn!

 **Check Your Score Online!**

*Visit **ViewMath Academy** to enter your answers and see which topics you need to review. You can also explore lessons, take quizzes, track your scores, and save your progress!*

viewmath.com/score/7.1.FL.17

Or go to viewmath.com/score and enter code: 7.1.FL.17

Practice Test 3

 30 Questions

✏ Before You Start ✏

- ✔ **Read each question carefully** before choosing your answer.
- ✔ **Show your work** on scratch paper when you need to.
- ✔ **Skip hard questions** and come back to them later.
- ✔ **Check your answers** when you're done.
- ✔ **Take your time** — there's no rush!

★ You've Got This! ★

Do your best and show what you know!

1. A proportional table has $k = \frac{3}{4}$. Which pair could be in the table?

(A) $(8, 10)$ (B) $(8, 6)$

(C) $(12, 8)$ (D) $(4, 2)$

2. A recipe for 4 people uses $\frac{3}{2}$ cups of rice. How much rice is needed for 10 people?

(A) $2\frac{1}{2}$ cups   (B) 3 cups

(C) $3\frac{3}{4}$ cups   (D) 5 cups

3. 63 is 90% of what number?

(A) 56.7   (B) 67

(C) 70   (D) 72

4. Sarah and Jake both solve "What is 40% of 75?" Sarah uses a proportion; Jake uses an equation. Which statement is true?

(A) Only the proportion method works   (B) Only the equation method works

(C) Both methods give the same answer   (D) The methods give different answers

5. What is the simple interest on $1,200 at 3.5% for 2 years?

(A) $42   (B) $72

(C) $84   (D) $210

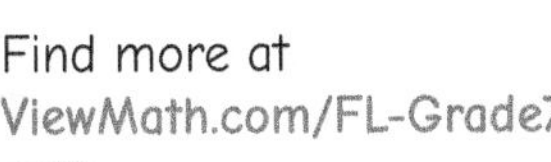

Find more at
ViewMath.com/FL-Grade7

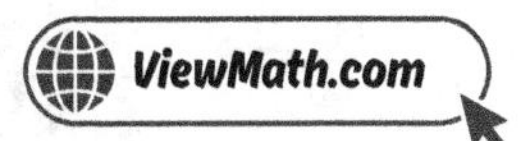

6. A student measured a table as 4.5 feet long. The actual length is 5 feet. What is the percent error?

(A) 0.5%

(B) 5%

(C) 10%

(D) 11.1%

7. A bank account has $-\$50$ (overdrawn). A deposit of $75 is made. What is the new balance?

(A) $-\$125$

(B) $-\$25$

(C) $\$25$

(D) $\$125$

8. Which expression is equivalent to $(-5) - 3$?

(A) $(-5) + 3$

(B) $(-5) + (-3)$

(C) $5 + 3$

(D) $5 + (-3)$

9. Factor $16x + 24y - 8$.

(A) $4(4x + 6y - 2)$

(B) $8(2x + 3y - 1)$

(C) $2(8x + 12y - 4)$

(D) $8(2x + 3y + 1)$

10. Simplify $(3.5n - 2.1) - (1.5n + 0.9)$.

(A) $2n - 3$

(B) $5n - 3$

(C) $2n - 1.2$

(D) $2n + 3$

11. A student solved $3(x + 4) = 30$ and wrote: $3x + 4 = 30$, $3x = 26$, $x = \frac{26}{3}$. What was the student's error?

(A) Divided by 3 instead of subtracting

(B) Did not distribute 3 to the 4

(C) Subtracted 4 from the wrong side

(D) Forgot to check the answer

Find more at
ViewMath.com/FL-Grade7

ViewMath.com

12. Solve $2.5x - 7.5 = 10$.

Your Answer:

13. Solve $-4x > 20$.

(A) $x > -5$

(B) $x < -5$

(C) $x > 5$

(D) $x < 5$

14. Solve $-5x + 15 \geq 30$. Describe the graph.

Your Answer:

15. Original scale: $1\ cm = 10\ m$. New scale: $1\ cm = 40\ m$. What is the scale ratio?

(A) $\frac{1}{4}$

(B) 4

(C) 30

(D) $\frac{1}{30}$

16. Two angles of a triangle are 25° and 25°. The triangle is:

(A) Equilateral

(B) Isosceles

(C) Scalene

(D) Impossible

Get Online

Find more at
ViewMath.com/FL-Grade7

ViewMath.com

17. The diagram shows a cone being sliced. Which cut produces a circular cross-section?

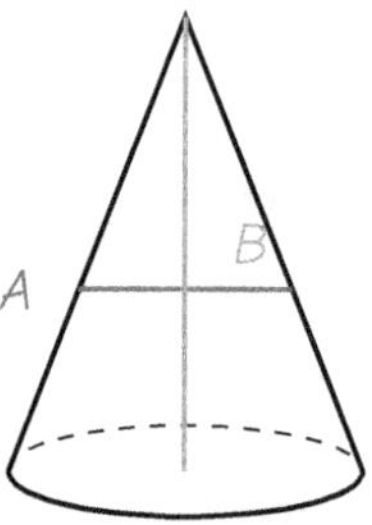

(A) Cut A (horizontal)

(B) Cut B (vertical through apex)

(C) Both cuts

(D) Neither cut

18. An angle is 4 times its supplement. What is the angle?

(A) 36°

(B) 45°

(C) 120°

(D) 144°

19. Which of the following is NOT a part of a circle?

(A) Center

(B) Radius

(C) Vertex

(D) Diameter

20. A cereal box is 30 cm tall, 20 cm wide, and 6 cm deep. What is its surface area?

(A) 1,440 cm^2

(B) 1,560 cm^2

(C) 1,800 cm^2

(D) 3,600 cm^2

21. A rectangular prism has a volume of 240 cm^3. Its base is 8 cm by 6 cm. What is the height?

(A) 4 cm

(B) 5 cm

(C) 6 cm

(D) 10 cm

Find more at
ViewMath.com/FL-Grade7

ViewMath.com

22. A cylindrical can is shown. Find the surface area of the can. Use $\pi \approx 3.14$.

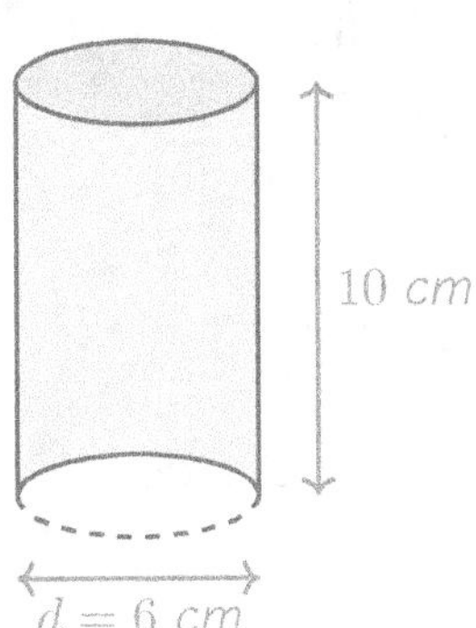

Your Answer:

23. A poll found that 35% of a random sample of 400 students prefer pizza for lunch. Which statement is the best inference?

(A) Exactly 35% of all students prefer pizza

(B) About 35% of the entire student population likely prefers pizza

(C) No students prefer anything other than pizza

(D) The poll is invalid because not all students were surveyed

24. When is a dot plot a better choice than a box plot for comparing two groups?

(A) When data sets are very large

(B) When you want to see every individual data point

(C) When you only want to compare medians

(D) When the data is categorical

25. Class A: median 85, IQR 8, range 20. Class B: median 80, IQR 15, range 40. A student says "Class B is better because it has a wider range." Is the student correct?

(A) Yes — wider range means more talent

(B) No — wider range means less consistency, and Class A has a higher median

(C) Yes — Class B's scores spread more, which is always better

(D) No — range does not measure anything useful

Find more at
ViewMath.com/FL-Grade7

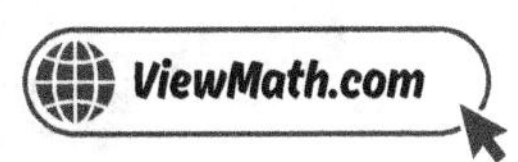

26. How many data values are shown in this stem-and-leaf plot?

Stem 5: 1 3 5   Stem 6: 0 2 4 7 8   Stem 7: 1 6   Stem 8: 0

Your Answer:

27. Two circle graphs compare how Class A and Class B spend their free time. Class A shows 45% on sports; Class B shows 30% on sports. If Class A has 20 students and Class B has 30 students, how many more students in Class B play sports than in Class A?

(A) 0 — the same number

(B) 9 is more in Class B

(C) 3 is more in Class A

(D) 1 more in Class A

28. A bag contains 4 red marbles and 1 blue marble. Which type of probability model does this represent?

(A) Uniform

(B) Non-uniform

(C) Both uniform and non-uniform

(D) Neither

29. Two dice are rolled. What is the probability of getting a sum of 2?

(A) $\frac{1}{6}$

(B) $\frac{1}{12}$

(C) $\frac{1}{36}$

(D) $\frac{2}{36}$

30. You simulate flipping a coin 3 times and counting how many times all 3 are heads. In 40 simulations, all 3 heads occurs 6 times. What is the experimental probability?

(A) $\frac{6}{40} = 0.15$

(B) $\frac{1}{8} = 0.125$

(C) $\frac{3}{40} = 0.075$

(D) $\frac{6}{120} = 0.05$

Find more at
ViewMath.com/FL-Grade7

ViewMath.com

 End of Practice Test 3 

Great job finishing the test!

☑ My Score

I got ____________ out of 30 questions right.

Check your answers in the **Answer Key** at the back of the book.

💡 Review any questions you missed. That's how we learn!

📊 Check Your Score Online!

Visit **ViewMath Academy** to enter your answers and see which topics you need to review. You can also explore lessons, take quizzes, track your scores, and save your progress!

viewmath.com/score/7.1.FL.18

Or go to viewmath.com/score and enter code: 7.1.FL.18

Practice Test 4

30 Questions

✏ Before You Start ✏

- ✔ **Read each question carefully** before choosing your answer.
- ✔ **Show your work** on scratch paper when you need to.
- ✔ **Skip hard questions** and come back to them later.
- ✔ **Check your answers** when you're done.
- ✔ **Take your time** — there's no rush!

★ You've Got This! ★

Do your best and show what you know!

1. What is the constant of proportionality (k) for the table below?

x	y
3	12
5	20
8	32

(A) 3

(B) 4

(C) 8

(D) 12

2. A truck delivers 540 packages in 3 trips. How many trips are needed to deliver 1,620 packages?

Your Answer:

3. 9 is 12% of what number?

(A) 1.08

(B) 36

(C) 72

(D) 75

4. A school garden has 80 plants. If 45% are tomato plants, how many tomato plants are there?

Your Answer:

5. Find the simple interest: $P = \$500$, $r = 6\%$, $t = 3$ years.

(A) $30

(B) $60

(C) $90

(D) $150

6. In the percent error formula, why do we use absolute value?

   (A) To make the calculation easier

   (B) Because the answer must be positive

   (C) Because we only care about the size of the error, not the direction

   (D) To convert the answer to a percent

7. What value of n makes $(-9) + n = -2$ true?

   (A) -11

   (B) -7

   (C) 7

   (D) 11

8. Find the distance between -11 and -3 on a number line.

   Your Answer:

9. Factor $-12y + 8$.

   (A) $4(-3y + 2)$

   (B) $-4(3y + 2)$

   (C) $-4(3y - 2)$

   (D) Both A and C

10. Simplify $(6a - 2) - (3a + 5)$.

    (A) $3a + 3$

    (B) $3a - 7$

    (C) $9a - 7$

    (D) $3a + 7$

11. Solve $7(x - 2) = 0$.

    (A) $x = 0$

    (B) $x = -2$

    (C) $x = 2$

    (D) $x = 7$

Find more at
ViewMath.com/FL-Grade7

ViewMath.com

12. Solve $\dfrac{x}{2} + \dfrac{x}{4} = 9$.

Your Answer:

13. Look at the number line. Which inequality has this solution?

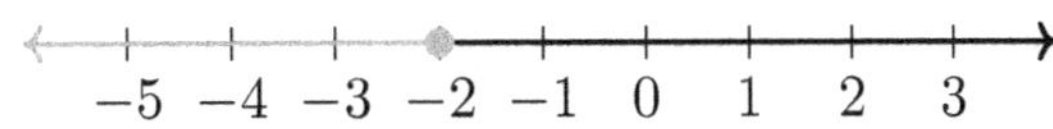

(A) $3x + 1 > -5$

(B) $-4x \geq 8$

(C) $2x - 3 < -7$

(D) $x + 5 > 3$

14. Solve $10 - 4x < 2$. Describe the graph.

Your Answer:

15. A map uses $1\ cm = 20\ km$. You redraw it at $1\ cm = 10\ km$. What happens to the drawing?

(A) Every length is halved

(B) Every length stays the same

(C) Every length is doubled

(D) Every length is multiplied by 10

Find more at
ViewMath.com/FL-Grade7

ViewMath.com

16. *The diagram shows two triangles. Both have angles 30°, 60°, and 90°, but different side lengths. What does this demonstrate?*

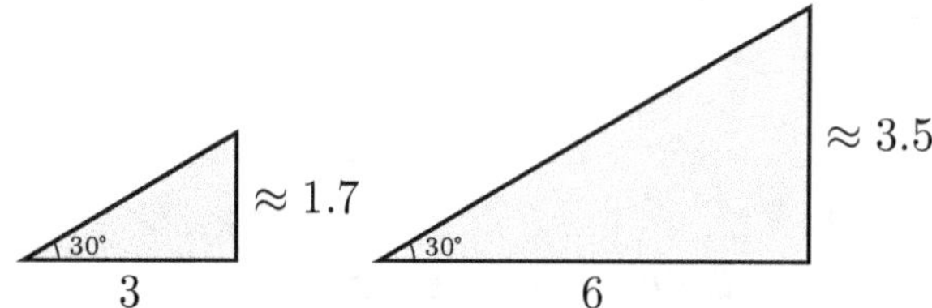

(A) SSS produces multiple triangles

(B) AAA produces a unique triangle

(C) AAA produces multiple triangles of different sizes

(D) These two triangles are congruent

17. *Which cross-section is NOT possible from slicing a rectangular prism?*

(A) Rectangle

(B) Triangle

(C) Circle

(D) Pentagon

18. *Two lines intersect. One angle formed is 130°. What are the measures of all four angles?*

(A) 130°, 130°, 50°, 50°

(B) 130°, 50°, 130°, 100°

(C) 130°, 130°, 130°, 130°

(D) 130°, 60°, 130°, 40°

19. *How many radii does it take to make one diameter?*

(A) 1

(B) 2

(C) 3

(D) 4

Find more at
ViewMath.com/FL-Grade7

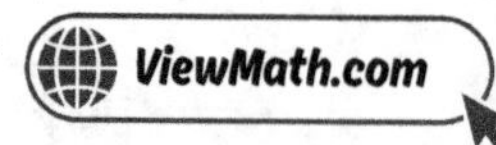

20. A rectangular prism has dimensions 7 m by 4 m by 3 m. What is the area of the largest face?

(A) $12\ m^2$

(B) $21\ m^2$

(C) $28\ m^2$

(D) $84\ m^2$

21. A rectangular fish tank is 50 cm long, 30 cm wide, and 40 cm tall. What is the volume?

(A) $120\ cm^3$

(B) $6{,}000\ cm^3$

(C) $60{,}000\ cm^3$

(D) $600\ cm^3$

22. What is the surface area of a cylinder with radius 3 cm and height 7 cm? Use $\pi \approx 3.14$.

(A) $131.88\ cm^2$

(B) $175.84\ cm^2$

(C) $188.4\ cm^2$

(D) $197.82\ cm^2$

23. The diagram below shows four different sampling methods used to survey students at a school. Which method is most likely to produce an unbiased sample?

Method A

Survey the football team

Method B

Survey every 10th student on the school roster

Method C

Post survey online and wait for responses

Method D

Survey students in one math class

(A) Method A

(B) Method B

(C) Method C

(D) Method D

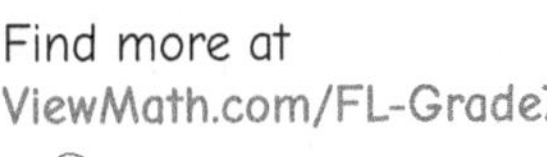

Find more at
ViewMath.com/FL-Grade7

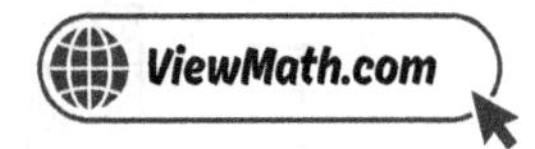
ViewMath.com

24. *Team A: mean 60, MAD 4. Team B: mean 70, MAD 3. Is the difference in means meaningful? Explain briefly.*

Your Answer:

25. *Group A has a mean of 50 and a MAD of 3. Group B has a mean of 60 and a MAD of 3. The difference in means expressed in MADs is:*

(A) $\frac{10}{3} \approx 3.3$ *MADs*

(B) *10 MADs*

(C) *3 MADs*

(D) *30 MADs*

26. *Data:* $45, 48, 52, 55, 55, 58, 61, 63$. *Which is the correct stem-and-leaf plot?*

(A) *Stems:* $4, 5, 6$; *Row 4:* 5 8; *Row 5:* 2 5 5 8; *Row 6:* 1 3

(B) *Stems:* $4, 5, 6$; *Row 4:* 5 8; *Row 5:* 5 5 2 8; *Row 6:* 3 1

(C) *Stems:* $45, 48, 52$; *no leaves*

(D) *Stems:* $4, 5$; *Row 4:* 5 8; *Row 5:* 2 5 5 8 1 3

27. *A student creates a circle graph with sectors of 100°, 80°, 60°, and 100°. Is this circle graph correct?*

(A) *Yes — the sectors look reasonable*

(B) *No — the angles add to 340°, not 360°*

(C) *No — there are too many sectors*

(D) *Yes — all angles are less than 180°*

28. *A coin is weighted so that heads comes up 60% of the time. What is the probability of tails?*

(A) 0.60

(B) 0.50

(C) 0.40

(D) 0.30

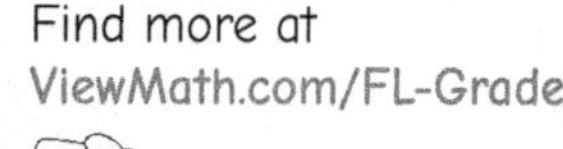

Find more at
ViewMath.com/FL-Grade7

ViewMath.com

29. Look at the tree diagram for flipping three coins. What is the probability of getting exactly 2 tails? Write your answer as a fraction in simplest form.

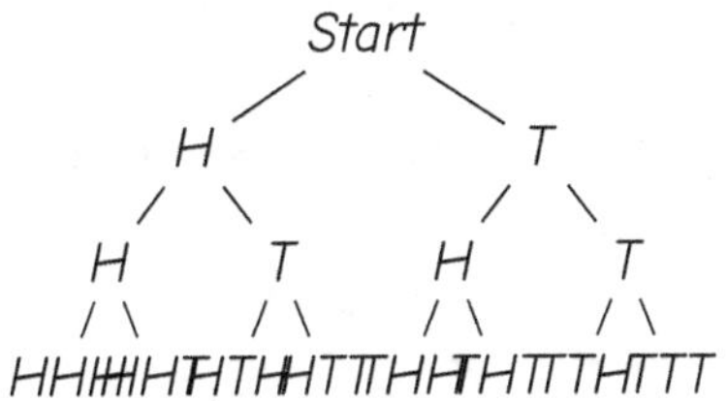

Your Answer:

30. A student wants to simulate the probability that exactly 2 out of 3 students pass a test, where each student has a 50% chance. Which model would work?

(A) Roll one die: odd = pass, even = fail

(B) Flip 3 coins: heads = pass, tails = fail; count exactly 2 heads

(C) Draw 3 marbles from a bag of 2 red and 1 blue

(D) Flip 1 coin 3 times and count all tails

 # End of Practice Test 4 

Great job finishing the test!

 My Score

I got ___________ out of 30 questions right.

*Check your answers in the **Answer Key** at the back of the book.*

💡 *Review any questions you missed. That's how we learn!*

Check Your Score Online!

Visit **ViewMath Academy** to enter your answers and see which topics you need to review. You can also explore lessons, take quizzes, track your scores, and save your progress!

viewmath.com/score/7.1.FL.19

Or go to viewmath.com/score and enter code: 7.1.FL.19

Practice Test 5

30 Questions

✏️ **Before You Start** ✏️

- ✔ **Read each question carefully** before choosing your answer.
- ✔ **Show your work** on scratch paper when you need to.
- ✔ **Skip hard questions** and come back to them later.
- ✔ **Check your answers** when you're done.
- ✔ **Take your time** — there's no rush!

⭐ You've Got This! ⭐

Do your best and show what you know!

1. A factory produces widgets at a constant rate. In $2\frac{1}{2}$ hours it makes 175 widgets. What is k in widgets per hour?

Your Answer:

2. A map scale says 2 cm = 15 miles. Two cities are 7 cm apart on the map. What is the actual distance?

(A) 42.5 miles

(B) 45 miles

(C) 52.5 miles

(D) 105 miles

3. A school survey showed that 54 out of 200 students walk to school. What percent of students walk to school?

(A) 25%

(B) 27%

(C) 37%

(D) 54%

4. Look at the proportion model below. What value of n makes the proportion true?

(A) 28

(B) 30

(C) 32

(D) 37.5

5. A savings account pays 3% simple interest. You deposit $2,000. After how many years will you earn $300 in interest?

(A) 3 years

(B) 4 years

(C) 5 years

(D) 10 years

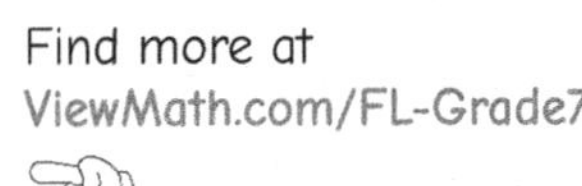
Find more at
ViewMath.com/FL-Grade7

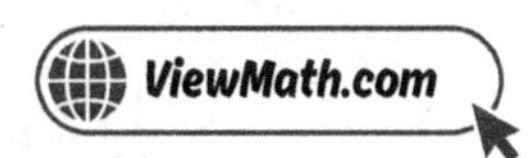
ViewMath.com

6. An architect estimated a building would be 45 meters tall. The actual height is 42 meters. What is the percent error (rounded to the nearest tenth)?

Your Answer:

7. What is $15 + (-15)$?

(A) 30

(B) -30

(C) 15

(D) 0

8. The high temperature was 4°F and the low was -11°F. What is the difference between the high and low?

(A) 7°F

(B) -7°F

(C) 15°F

(D) -15°F

9. Factor $-15a + 25$.

(A) $5(-3a + 5)$

(B) $-5(3a + 5)$

(C) $-5(3a - 5)$

(D) Both A and C

10. The total cost for two items is $(3x + 7) + (5x - 2)$. A coupon saves you $(2x + 1)$. What do you pay after the coupon?

Your Answer:

11. Solve $4(x + 9) = 52$.

Your Answer:

Find more at
ViewMath.com/FL-Grade7

ViewMath.com

12. Which is the best first step to solve $\dfrac{x}{4} + \dfrac{1}{2} = 3$?

   (A) Subtract $\frac{1}{2}$ from both sides   (B) Multiply every term by 4

   (C) Divide both sides by 4   (D) Add $\frac{1}{2}$ to both sides

13. Solve $8 - 3x \geq 23$.

Your Answer

14. Which value is NOT a solution to $x \geq -4$?

   (A) $x = -4$ (B) $x = 0$

   (C) $x = -5$ (D) $x = 100$

15. A floor plan uses $1\ cm = 5\ ft$. You redraw at $1\ cm = 2\ ft$. A room that is 3 cm by 4 cm on the original drawing. What is the area on the new drawing?

   (A) $12\ cm^2$ (B) $30\ cm^2$

   (C) $75\ cm^2$ (D) $150\ cm^2$

16. What condition (SSS, SAS, ASA, AAS, or AAA) gives infinitely many triangles?

Your Answer

17. Which statement about cross-sections is true?

    (A) The cross-section of any 3D figure is always a circle

    (B) The cross-section depends on the angle and position of the cut

    (C) A cross-section is always the same shape as the base

    (D) Only prisms can have cross-sections

18. Two vertical angles are formed by intersecting lines. One angle is 72°. What is the vertical angle?

    (A) 18°

    (B) 72°

    (C) 108°

    (D) 288°

19. Which part of a circle is labeled O in most diagrams?

    (A) A point on the circumference

    (B) The center

    (C) The diameter

    (D) The radius

20. A rectangular prism is 6 ft by 6 ft by 10 ft. What is its surface area?

    (A) 72 ft^2

    (B) 240 ft^2

    (C) 312 ft^2

    (D) 360 ft^2

21. A rectangular aquarium is 80 cm long, 40 cm wide, and 50 cm tall. What is its volume in cubic centimeters?

    Your Answer:

22. What is the surface area of a cylinder with radius 5 cm and height 8 cm? Use $\pi \approx 3.14$.

    (A) 251.2 cm^2

    (B) 314 cm^2

    (C) 408.2 cm^2

    (D) 628 cm^2

Find more at
ViewMath.com/FL-Grade7

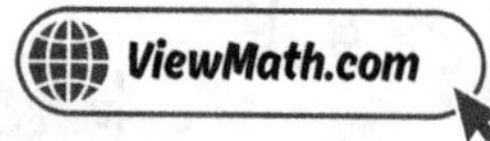

23. A doctor wants to study the effect of a new medicine on 10,000 patients. She randomly selects 500 patients for a trial. What is the sample size?

(A) 10,000

(B) 500

(C) 10,500

(D) 50

24. Class A has a mean test score of 78 and Class B has a mean test score of 85. What can you conclude?

(A) Every student in Class B scored higher than every student in Class A

(B) Class B scored higher on average than Class A

(C) Class A and Class B have the same scores

(D) Class A has more students than Class B

25. Team X: mean 82, MAD 5. Team Y: mean 76, MAD 5. Which team performed better on average, and is the difference meaningful?

(A) Team X; no, because the difference is less than 2 MADs

(B) Team X; yes, because $\frac{6}{5} = 1.2$ MADs, which is meaningful

(C) Team X; yes, because the means are more than 2 MADs apart

(D) Team Y; yes, because its MAD is the same

26. A stem-and-leaf plot shows the row 2 | 2 2 5 8. What is the mode for values with stem 2?

(A) 22

(B) 25

(C) 28

(D) There is no mode

27. A sector represents 15% of a data set. What is its angle in degrees?

Your Answer:

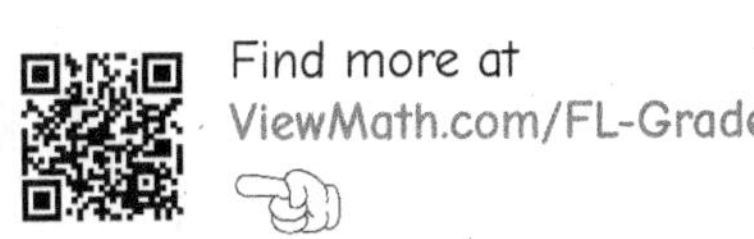

Find more at
ViewMath.com/FL-Grade7

28. A bag has tiles labeled A, A, A, B, B. What is the probability of drawing a B?

(A) $\frac{1}{5}$

(B) $\frac{2}{3}$

(C) $\frac{2}{5}$

(D) $\frac{3}{5}$

29. The table shows all possible products when two dice are rolled. How many outcomes give a product of 6? Write the probability as a fraction in simplest form.

×	1	2	3	4	5	6
1	1	2	3	4	5	6
2	2	4	6	8	10	12
3	3	6	9	12	15	18
4	4	8	12	16	20	24
5	5	10	15	20	25	30
6	6	12	18	24	30	36

Your Answer:

30. Which of the following is the FIRST step in designing a simulation?

(A) Record the results

(B) Run many trials

(C) Identify the event and its possible outcomes

(D) Calculate the experimental probability

Find more at
ViewMath.com/FL-Grade7

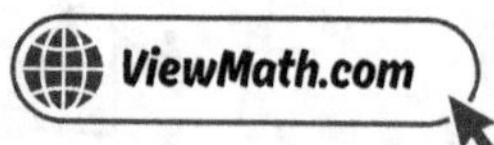

 # End of Practice Test 5 

Great job finishing the test!

 My Score

I got _____________ out of 30 questions right.

*Check your answers in the **Answer Key** at the back of the book.*

Review any questions you missed. That's how we learn!

 Check Your Score Online!

Visit **ViewMath Academy** to enter your answers and see which topics you need to review. You can also explore lessons, take quizzes, track your scores, and save your progress!

viewmath.com/score/7.1.FL.20

Or go to viewmath.com/score and enter code: 7.1.FL.20

Practice Test 6

30 Questions

✏ Before You Start ✏

- ✔ **Read each question carefully** before choosing your answer.
- ✔ **Show your work** on scratch paper when you need to.
- ✔ **Skip hard questions** and come back to them later.
- ✔ **Check your answers** when you're done.
- ✔ **Take your time** — there's no rush!

★ You've Got This! ★

Do your best and show what you know!

15. A blueprint uses $1\ in = 12\ ft$. You redraw at $1\ in = 4\ ft$. A door that is $0.5\ in$ wide on the original becomes how wide?

(A) $0.5\ in$

(B) $1\ in$

(C) $1.5\ in$

(D) $3\ in$

16. Can a triangle have angles $90°$, $100°$, and $-10°$?

(A) Yes, it is a right triangle

(B) Yes, because the sum is $180°$

(C) No, because an angle cannot be negative

(D) No, because the sum is not $180°$

17. A rectangular prism has a base of $6\ cm$ by $4\ cm$ and a height of $10\ cm$. It is sliced parallel to the base. What are the dimensions of the cross-section?

Your Answer

18. Three angles meet at a point on a line: $50°$, $(2x + 10)°$, and $x°$. Find x.

Your Answer:

19. Which statement about the diameter and radius of a circle is always true?

(A) The diameter is half the radius

(B) The diameter equals the radius

(C) The diameter is twice the radius

(D) The diameter is three times the radius

20. A rectangular prism has dimensions $4\ m$, $4\ m$, and $10\ m$. How much greater is its surface area than a cube with side $4\ m$?

Your Answer

Find more at
ViewMath.com/FL-Grade7

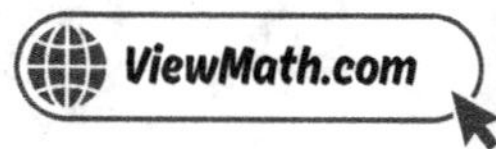

11. Solve $10(x - 5) = 30$.

(A) $x = 8$

(B) $x = 3$

(C) $x = -2$

(D) $x = 35$

12. The table shows input and output values for a function. What is the rule?

Input (x)	Output (y)
2	2.5
4	3.5
6	4.5
10	6.5

(A) $y = 0.5x + 1.5$

(B) $y = x - 0.5$

(C) $y = 0.25x + 2$

(D) $y = 2x - 1.5$

13. Solve $-7x - 4 < 17$.

Your Answer

14. A parking garage charges \$5 per hour. You have \$30. Which inequality represents the hours h you can park, and what does the graph look like?

(A) $5h \leq 30$; closed circle at 6, shade left from 0 to 6

(B) $5h < 30$; open circle at 6, shade left

(C) $5h \geq 30$; closed circle at 6, shade right

(D) $5h > 30$; open circle at 6, shade right

Get Online

Find more at
ViewMath.com/FL-Grade7

ViewMath.com

6. A scientist predicted a chemical reaction would produce 30 grams. It actually produced 28 grams. What is the percent error?

(A) 2%

(B) 6.7%

(C) 7.1%

(D) 14.3%

7. Which expression has a positive sum?

(A) $(-8) + 3$

(B) $(-5) + (-2)$

(C) $(-4) + 9$

(D) $(-10) + 6$

8. If $a = -8$ and $b = 5$, what is $a - b$?

(A) -3

(B) 3

(C) -13

(D) 13

9. Factor $15a - 10$.

(A) $5(3a - 2)$

(B) $5(3a + 2)$

(C) $3(5a - 10)$

(D) $10(a - 1)$

10. Simplify $(2y + 7) - (5y - 3)$.

(A) $-3y + 10$

(B) $-3y + 4$

(C) $7y + 10$

(D) $3y + 10$

Get Online

Find more at
ViewMath.com/FL-Grade7

ViewMath.com

1. A proportional graph passes through $(6, 15)$. What is k?

Your Answer:

2. A printer prints 120 pages in 8 minutes. How long will it take to print 450 pages?

(A) 25 min

(B) 28 min

(C) 30 min

(D) 36 min

3. A library has 480 books. If 35% are fiction, how many fiction books does the library have?

(A) 148

(B) 160

(C) 168

(D) 172

4. 24 out of 80 students chose pizza for lunch. What percent is that?

(A) 24%

(B) 28%

(C) 30%

(D) 32%

5. Two bank accounts both start with $1,000. Account A earns 5% for 2 years. Account B earns 2% for 5 years. Which earns more interest?

(A) Account A

(B) Account B

(C) They earn the same

(D) Cannot be determined

Find more at
ViewMath.com/FL-Grade7

 End of Practice Test 8 

Great job finishing the test!

 My Score

I got _____________ out of 30 questions right.

*Check your answers in the **Answer Key** at the back of the book.*

💡 *Review any questions you missed. That's how we learn!*

📊 Check Your Score Online!

Visit **ViewMath Academy** to enter your answers and see which topics you need to review. You can also explore lessons, take quizzes, track your scores, and save your progress!

viewmath.com/score/7.1.FL.23

Or go to viewmath.com/score and enter code: 7.1.FL.23

29. A coin is flipped and a die is rolled. What is the probability of getting tails and a 1? Write your answer as a fraction.

Your Answer:

30. How does increasing the number of trials in a simulation affect the results?

(A) The results become less reliable

(B) The experimental probability gets closer to the theoretical probability

(C) The results stay exactly the same

(D) The simulation takes less time

Find more at
ViewMath.com/FL-Grade7

25. *Group X: mean 45, MAD 6. Group Y: mean 57, MAD 6. Express the difference in means as a multiple of MAD.*

Your Answer:

26. How many total data values are in this stem-and-leaf plot?

Stem	Leaf
4	2 5 8
5	0 3 3 6 9
6	1 4

(A) 3

(B) 8

(C) 10

(D) 15

27. *A survey of 60 people: 18 chose option A, 24 chose option B, 18 chose option C. Find the angle for option B in a circle graph.*

Your Answer:

28. *A bag contains 5 yellow and 5 purple marbles. Lisa picks a marble, records the color, and replaces it 40 times. She gets yellow 24 times. Which statement is correct?*

(A) The model predicts $P(\text{yellow}) = 0.5$, and the experiment gives $P(\text{yellow}) = 0.6$, so there is a small difference

(B) The model is wrong because $24 \neq 20$

(C) The model predicts $P(\text{yellow}) = 0.6$

(D) The experiment proves yellow is more likely than purple

Find more at
ViewMath.com/FL-Grade7

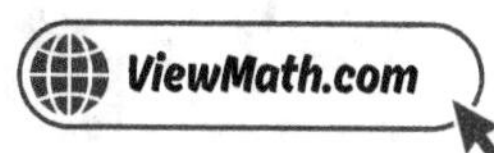

20. The diagram shows a triangular prism. What is its total surface area?

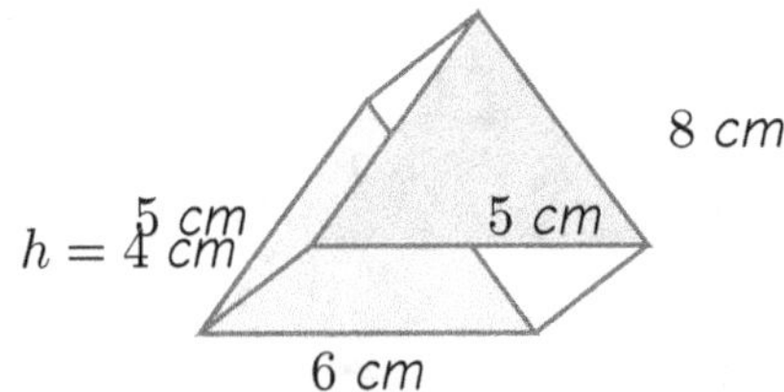

(A) 104 cm^2

(B) 128 cm^2

(C) 152 cm^2

(D) 160 cm^2

21. A rectangular prism has a volume of 360 cm^3. Its length is 12 cm and its height is 5 cm. What is its width?

Your Answer:

22. Which formula gives the volume of a cylinder?

(A) $V = 2\pi rh$

(B) $V = \pi r^2 h$

(C) $V = \pi r^2$

(D) $V = \frac{1}{3}\pi r^2 h$

23. Which sampling method is most likely to produce a representative sample of a school's students?

(A) Survey students in the cafeteria during lunch

(B) Survey every fifth student on an alphabetical class list

(C) Survey students who volunteer to participate

(D) Survey only students in honors classes

24. Data set: 4, 6, 7, 8, 10. The mean is 7. Find the MAD.

Your Answer:

Get Online

Find more at
ViewMath.com/FL-Grade7

ViewMath.com

15. A floor plan uses $1 \text{ cm} = 6 \text{ ft}$. On this plan, a room is 9 cm long. On a second plan, the same room is 3 cm long. What scale does the second plan use?

Your Answer:

16. A triangle has angles 60°, 60°, and 60°, and one side measures 10 cm. How many triangles can be drawn?

Your Answer:

17. What shape is the cross-section when a cone is sliced vertically through its apex?

Your Answer:

18. Which statement about vertical angles is always true?

(A) They are supplementary

(B) They are complementary

(C) They are equal

(D) They sum to 360°

19. The circumference of a circle is 62.8 cm. Using $\pi \approx 3.14$, find the radius.

Your Answer:

Find more at
ViewMath.com/FL-Grade7

ViewMath.com

11. *The diagram shows two equal groups. Which equation represents this, and what is x?*

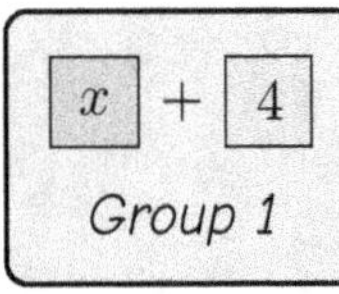

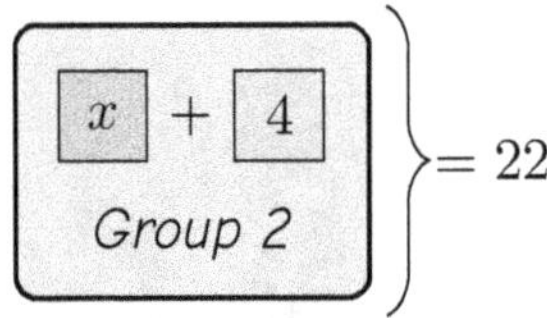

(A) $2(x + 4) = 22$, $x = 7$

(B) $2x + 4 = 22$, $x = 9$

(C) $x + 8 = 22$, $x = 14$

(D) $2(x + 4) = 22$, $x = 3$

12. *Solve* $\dfrac{x}{6} - \dfrac{2}{3} = \dfrac{1}{2}$.

(A) $x = 3$

(B) $x = 7$

(C) $x = -1$

(D) $x = 10$

13. *Solve* $\dfrac{x}{2} + 7 \le 12$.

Your Answer:

14. *A student has test scores of 78, 85, and 90. She needs an average of at least 85 on four tests to earn an A. Solve the inequality to find the minimum score s on the fourth test, and describe the graph.*

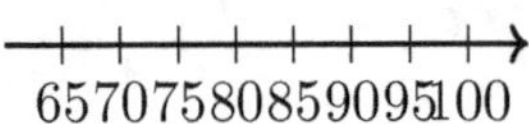

Your Answer:

Get Online

Find more at
ViewMath.com/FL-Grade7

ViewMath.com

6. The number line below shows an estimated value and an actual value. What is the percent error?

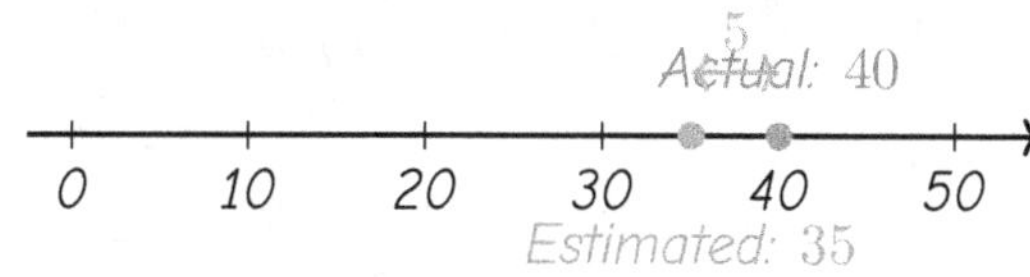

(A) 5%

(B) 10%

(C) 12.5%

(D) 14.3%

7. What is $(-25) + 13 + (-8)$?

(A) -20

(B) -46

(C) 30

(D) -4

8. Death Valley's lowest point is -282 feet. A nearby hill is 134 feet above sea level. What is the difference in elevation?

(A) 148 feet

(B) -148 feet

(C) 416 feet

(D) -416 feet

9. Factor $24x - 16$.

(A) $4(6x - 4)$

(B) $8(3x - 2)$

(C) $2(12x - 8)$

(D) $6(4x - 2)$

10. Simplify $(2x + 3) + (4x - 1) - (x + 5)$.

Your Answer

Get Online

Find more at
ViewMath.com/FL-Grade7

ViewMath.com

1. A car travels at a constant speed. It goes 150 miles in 2.5 hours. What is the constant of proportionality?

(A) 30 mph

(B) 60 mph

(C) 75 mph

(D) 150 mph

2. The table shows how many batches of muffins a bakery can make and the total flour used. How many cups of flour are needed for 15 batches?

Batches	Flour (cups)
3	$7\frac{1}{2}$
6	15
9	$22\frac{1}{2}$
15	?

Your Answer:

3. 56 is 80% of what number?

Your Answer:

4. 27 out of 180 students earned an A on the exam. What percent earned an A?

(A) 12%

(B) 15%

(C) 18%

(D) 27%

5. You earned $180 in interest on $2,000 over 3 years. What was the interest rate?

Your Answer:

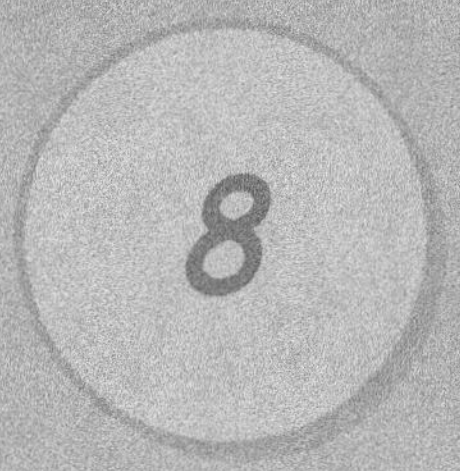

Practice Test 8

 30 Questions

 Before You Start 

- ✓ **Read each question carefully** before choosing your answer.
- ✓ **Show your work** on scratch paper when you need to.
- ✓ **Skip hard questions** and come back to them later.
- ✓ **Check your answers** when you're done.
- ✓ **Take your time** — there's no rush!

⭐ You've Got This! ⭐

Do your best and show what you know!

 **End of Practice Test 7** 

Great job finishing the test!

☑ My Score

I got _____________ out of 30 questions right.

*Check your answers in the **Answer Key** at the back of the book.*

💡 *Review any questions you missed. That's how we learn!*

📊 Check Your Score Online!

Visit **ViewMath Academy** to enter your answers and see which topics you need to review. You can also explore lessons, take quizzes, track your scores, and save your progress!

viewmath.com/score/7.1.FL.22

Or go to viewmath.com/score and enter code: 7.1.FL.22

27. A survey of 120 students asked about their favorite after-school activity. The results are:

Activity	Students
Sports	48
Music	30
Art	18
Reading	24

Find the percent and angle (in degrees) for each category.

Your Answer:

28. Which of the following is an example of a uniform probability model?

(A) A bag with 3 red, 5 blue, and 2 green marbles

(B) A fair number cube with 6 sides

(C) A spinner with sections of different sizes

(D) A weighted coin

29. A coin is flipped and a die is rolled. What is the probability of getting heads and a number greater than 4?

(A) $\frac{1}{12}$

(B) $\frac{1}{6}$

(C) $\frac{2}{12}$

(D) $\frac{1}{3}$

30. A game has a 25% chance of winning. Which simulation model correctly represents this?

(A) Flip a coin — heads means win

(B) Roll a die — roll a 1 means win

(C) Use a spinner with 4 equal sections — one section means win

(D) Use digits 0–9 — digits 0–4 mean win

Find more at
ViewMath.com/FL-Grade7

ViewMath.com

23. A random sample of 80 out of 2,000 students found that 20 students ride the bus. Predict how many students in the school ride the bus.

Your Answer:

24. The dot plots below show the number of push-ups completed by students in two gym classes.

Class 2

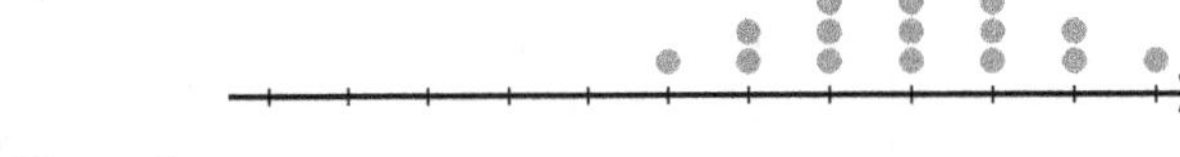

Class 1

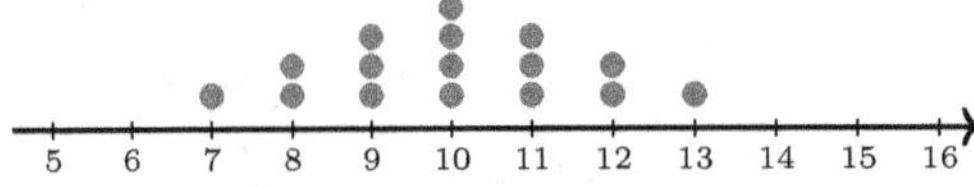

Which class has a higher center, and by approximately how many push-ups?

(A) Class 1, by about 3

(B) Class 2, by about 3

(C) Class 2, by about 6

(D) They have the same center

25. Explain one advantage of using IQR instead of range to describe the spread of a data set.

Your Answer:

26. Why is it important to include a key (e.g., "3 | 4 means 34") in a stem-and-leaf plot?

(A) It makes the plot look nicer

(B) It tells the reader how to interpret the stems and leaves

(C) It is not important

(D) It replaces the title

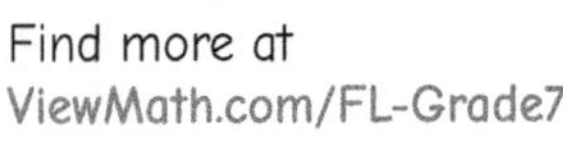

Find more at
ViewMath.com/FL-Grade7

ViewMath.com

21. A rectangular prism is 8 in by 5 in by 4 in. Another is 10 in by 4 in by 4 in. Which has more volume and by how much?

(A) First prism, by 10 in^3

(B) Second prism, by 10 in^3

(C) First prism, by 20 in^3

(D) They have the same volume

22. Find the surface area of a cylinder with radius 4 cm and height 9 cm. Use $\pi \approx 3.14$.

Your Answer:

23. A website asks visitors to rate a product. Only 5% of visitors respond. This is a:

(A) Random sample

(B) Systematic sample

(C) Voluntary response sample

(D) Stratified sample

24. Which display is best for comparing the overall shapes of two large data sets?

(A) A single dot plot

(B) Two histograms

(C) A circle graph

(D) A frequency table with one column

25. Two groups have the same mean but different MADs. What does this tell you?

(A) One group scored higher than the other

(B) The groups have different levels of variability

(C) The groups are identical

(D) The means were calculated incorrectly

26. Which data display keeps the original data values while also showing the shape of the data?

(A) Histogram

(B) Box plot

(C) Stem-and-leaf plot

(D) Circle graph

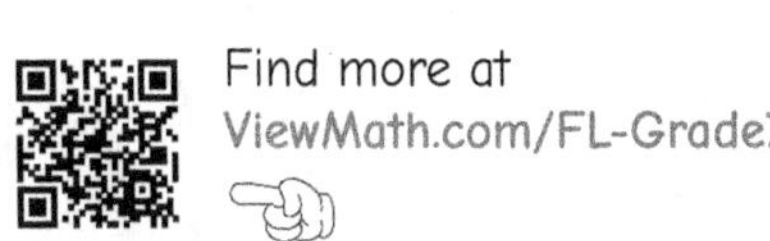
Find more at
ViewMath.com/FL-Grade7

27. A survey of 500 people found: Walking 20%, Driving 55%, Bus 15%, Other 10%. How many drive?

(A) 55

(B) 100

(C) 200

(D) 275

28. A spinner is divided into 4 sections with probabilities: red $= 0.4$, blue $= 0.3$, green $= 0.2$, yellow $= 0.1$. What type of model is this?

(A) Uniform, because there are 4 sections

(B) Non-uniform, because the probabilities are different

(C) Invalid, because the probabilities don't add up to 1

(D) Uniform, because all sections are on the same spinner

29. A coin is flipped and a die is rolled. What is the probability of getting tails and an even number?

(A) $\frac{1}{12}$

(B) $\frac{1}{6}$

(C) $\frac{1}{4}$

(D) $\frac{1}{2}$

30. What is a simulation?

(A) A way to calculate exact probabilities

(B) A model that uses random numbers or objects to imitate a real event

(C) A method that always gives the same result

(D) A graph that shows all possible outcomes

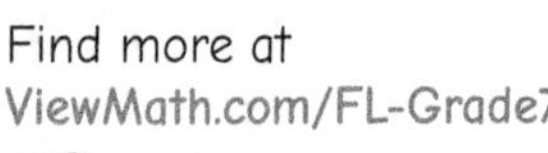

Find more at
ViewMath.com/FL-Grade7

 # End of Practice Test 6 

Great job finishing the test!

 **My Score**

I got ____________ out of 30 questions right.

*Check your answers in the **Answer Key** at the back of the book.*

Review any questions you missed. That's how we learn!

Check Your Score Online!

*Visit **ViewMath Academy** to enter your answers and see which topics you need to review. You can also explore lessons, take quizzes, track your scores, and save your progress!*

viewmath.com/score/7.1.FL.21

Or go to viewmath.com/score and enter code: 7.1.FL.21

Practice Test 7

📋 30 Questions

✏️ Before You Start ✏️

- ✔ **Read each question carefully** before choosing your answer.
- ✔ **Show your work** on scratch paper when you need to.
- ✔ **Skip hard questions** and come back to them later.
- ✔ **Check your answers** when you're done.
- ✔ **Take your time** — there's no rush!

★ You've Got This! ★

Do your best and show what you know!

1. *Find the missing value in the proportional table.*

x	y
5	20
9	?

Your Answer:

2. *A machine fills 350 bottles in 5 hours. How many bottles does it fill in 12 hours?*

(A) 700

(B) 780

(C) 840

(D) 900

3. *42 is what percent of 168?*

Your Answer:

4. *135 is 54% of what number?*

(A) 72.9

(B) 200

(C) 245

(D) 250

5. *A loan of $5,000 at 3% per year charges $450 in interest. How many years was the loan?*

Your Answer:

Find more at
ViewMath.com/FL-Grade7

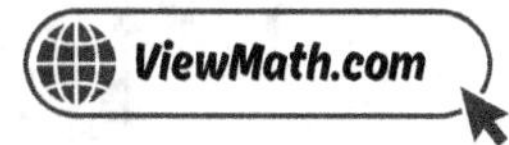

6. You estimated there were 50 marbles in a jar. The actual count was 40. What is the percent error?

(A) 10%   (B) 20%

(C) 25%   (D) 50%

7. What integer must you add to -15 to get -6?

Your Answer:

8. Which problem has the greatest answer?

(A) $5 - 12$ (B) $(-3) - (-10)$

(C) $(-8) - 1$ (D) $2 - 7$

9. What is the greatest common factor (GCF) of 12 and 18?

(A) 2   (B) 3

(C) 6   (D) 9

10. What is $(4x + 9) - (4x + 9)$?

(A) $8x + 18$ (B) $8x$

(C) 0   (D) 18

11. Each side of an equilateral triangle is $(2x + 5)$ cm. The perimeter is 57 cm. Find x.

Your Answer:

Find more at
ViewMath.com/FL-Grade7

ViewMath.com

12. Solve $-1.5x + 4.5 = -3$.

(A) $x = -5$

(B) $x = 1$

(C) $x = 5$

(D) $x = -1$

13. Solve $\dfrac{x}{3} + 4 < 10$.

(A) $x < 18$

(B) $x < 42$

(C) $x < 2$

(D) $x < 6$

14. Which of the following is the correct way to read the graph: open circle at 0, shading to the left?

(A) All positive numbers

(B) All negative numbers

(C) All numbers less than 0

(D) All numbers less than or equal to 0

15. A drawing uses 1 cm = 12 m. You redraw at 1 cm = 4 m. A hallway is 5 cm on the original. How long is it on the new drawing?

Your Answer:

16. A triangle has angles 40°, 60°, and 80°. How many different triangles can be drawn?

(A) None

(B) Exactly one

(C) Exactly three

(D) Infinitely many

17. You slice a rectangular prism with a cut parallel to its base. What shape is the cross-section?

(A) Triangle

(B) Circle

(C) Rectangle

(D) Pentagon

Find more at
ViewMath.com/FL-Grade7

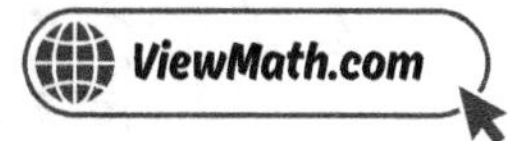

18. Two supplementary angles are equal. What is the measure of each?

(A) 45°

(B) 60°

(C) 90°

(D) 180°

19. The diameter of a bicycle wheel is 26 inches. What is the radius of the wheel?

(A) 52 inches

(B) 26 inches

(C) 13 inches

(D) 6.5 inches

20. A cube has a surface area of 384 in². What is the side length?

Your Answer:

21. A triangular prism has a triangular base with legs 5 cm and 12 cm (right triangle). The prism is 15 cm long. What is the volume?

(A) 150 cm³

(B) 300 cm³

(C) 450 cm³

(D) 900 cm³

22. A cylindrical water tank has a radius of 2 m and a height of 5 m. How many cubic meters of water can it hold? Use $\pi \approx 3.14$.

(A) 31.4 m³

(B) 62.8 m³

(C) 125.6 m³

(D) 20 m³

Find more at
ViewMath.com/FL-Grade7

9

Practice Test 9

 30 Questions

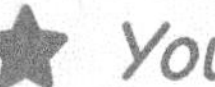 Before You Start 

- ✓ **Read each question carefully** before choosing your answer.
- ✓ **Show your work** on scratch paper when you need to.
- ✓ **Skip hard questions** and come back to them later.
- ✓ **Check your answers** when you're done.
- ✓ **Take your time** — there's no rush!

★ You've Got This! ★

Do your best and show what you know!

1. Two students find k from the point $(8, 20)$. Student A says $k = 2.5$. Student B says $k = 0.4$. Who is correct?

(A) Student A, because $k = \frac{y}{x}$

(B) Student B, because $k = \frac{x}{y}$

(C) Both are correct, depending on which variable is independent

(D) Neither is correct

2. Five T-shirts cost $42.50. At this rate, how much do 8 T-shirts cost?

(A) $56.00

(B) $64.00

(C) $68.00

(D) $72.00

3. What is 40% of 150?

(A) 45

(B) 50

(C) 55

(D) 60

4. In a bag of 150 marbles, 60 are blue. What percent of the marbles are blue?

(A) 25%

(B) 30%

(C) 35%

(D) 40%

5. Maria earned $120 in simple interest on a $1,500 deposit at 4%. How long was the money deposited?

(A) 1 year

(B) 2 years

(C) 3 years

(D) 4 years

Find more at
ViewMath.com/FL-Grade7

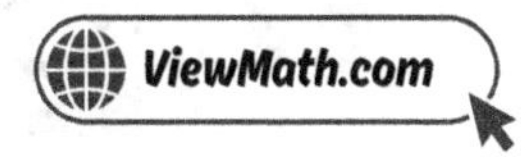

6. A builder estimated a project would cost $8,000. The actual cost was $7,500. What is the percent error?

  (A) 5%   (B) 6.25%

  (C) 6.67%   (D) 10%

7. You owe $25 and borrow $10 more. Which expression represents your total debt?

  (A) $25 + 10 = 35$ (B) $(-25) + 10 = -15$

  (C) $(-25) + (-10) = -35$ (D) $25 - 10 = 15$

8. The diagram shows two floors in a building. The parking garage is at level -3 and the rooftop is at level 5.

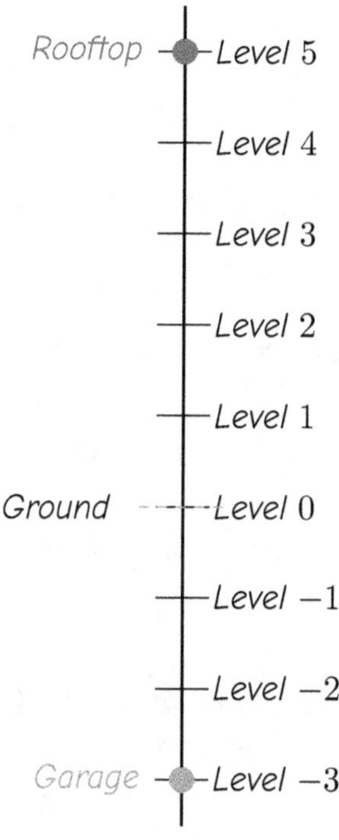

How many levels apart are the garage and the rooftop?

Your Answer

Get Online

Find more at
ViewMath.com/FL-Grade7

ViewMath.com

9. Which expression cannot be factored (has GCF of 1)?

(A) $6x + 9$

(B) $4a - 10$

(C) $5n + 7$

(D) $12y + 8$

10. A fence has three sections with the lengths shown. Find the total length, simplified.

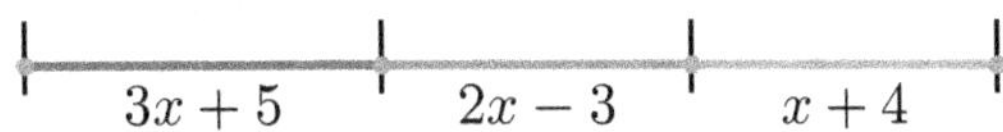

Your Answer:

11. Solve $5(n + 3) = -10$.

(A) $n = -5$

(B) $n = 1$

(C) $n = -1$

(D) $n = -\dfrac{25}{5}$

12. Solve $0.5x + 3 = 8$.

(A) $x = 10$

(B) $x = 22$

(C) $x = 5.5$

(D) $x = 2.5$

13. Which value of x is a solution to $4x - 1 > 11$?

(A) $x = 2$

(B) $x = 3$

(C) $x = 4$

(D) $x = 1$

Find more at
ViewMath.com/FL-Grade7

ViewMath.com

14. *Which graph represents $x \leq -2$?*

(A) *Open circle at −2, shade left*   (B) *Closed circle at −2, shade right*

(C) *Closed circle at −2, shade left*   (D) *Open circle at −2, shade right*

15. *A model uses scale 1 : 25. A new model uses scale 1 : 75. A window that is 6 cm on the first model becomes how long on the second?*

(A) 2 cm   (B) 6 cm

(C) 18 cm   (D) 150 cm

16. *Can a triangle be formed with sides 7, 10, and 15?*

(A) *No, the triangle inequality fails*   (B) *Yes, exactly one triangle*

(C) *Yes, more than one triangle*   (D) *Yes, but only a right triangle*

17. *A triangular prism is sliced with a cut parallel to its triangular base. What is the cross-section?*

(A) *A rectangle*   (B) *A triangle congruent to the base*

(C) *A smaller triangle*   (D) *A trapezoid*

18. *Two supplementary angles are $(2x + 10)°$ and $(3x)°$. What is the value of x?*

(A) 30   (B) 34

(C) 36   (D) 40

Find more at
ViewMath.com/FL-Grade7

19. A line segment that passes through the center of a circle and has both endpoints on the circle is called a
_________.

(A) Radius

(B) Chord

(C) Diameter

(D) Circumference

20. A triangular prism has two triangular bases each with area 15 cm^2. The three rectangular lateral faces have
areas 40 cm^2, 30 cm^2, and 50 cm^2. What is the total surface area?

(A) 120 cm^2

(B) 135 cm^2

(C) 150 cm^2

(D) 165 cm^2

21. A trapezoidal prism has a trapezoidal base with parallel sides 4 cm and 6 cm, height 3 cm, and the prism length
is 10 cm. What is its volume?

(A) 60 cm^3

(B) 120 cm^3

(C) 150 cm^3

(D) 180 cm^3

22. What is the volume of a cylinder with radius 3 cm and height 10 cm? Use $\pi \approx 3.14$.

(A) 30 cm^3

(B) 94.2 cm^3

(C) 282.6 cm^3

(D) 565.2 cm^3

23. Which question would require sampling rather than surveying the entire population?

(A) How many students are in a classroom of 25?

(B) What is the average lifespan of a light bulb produced by a factory?

(C) What is the total number of books on a shelf?

(D) How many chairs are in a room?

Find more at
ViewMath.com/FL-Grade7

24. Two groups have overlapping dot plots. What does the overlap indicate?

(A) The groups are identical

(B) Some data values from both groups are similar

(C) One group is always better than the other

(D) The data is incorrect

25. The table shows statistics for two basketball players' points per game over a season.

	Player A	Player B
Mean	18	20
Median	17	19
Range	10	25
MAD	3	7

Which player is more consistent?

(A) Player A, because the MAD and range are both smaller

(B) Player B, because the mean is higher

(C) Player A, because the mean is lower

(D) Player B, because the range is larger

26. A stem-and-leaf plot shows 12 data values. What is the median?

(A) The 6th value

(B) The 7th value

(C) The average of the 6th and 7th values

(D) The 12th value

Find more at
ViewMath.com/FL-Grade7

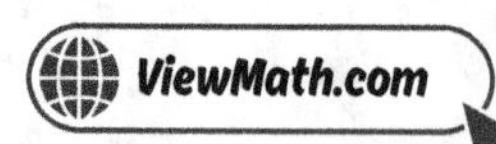

27. The circle graph shows how a student spends a 24-hour day.

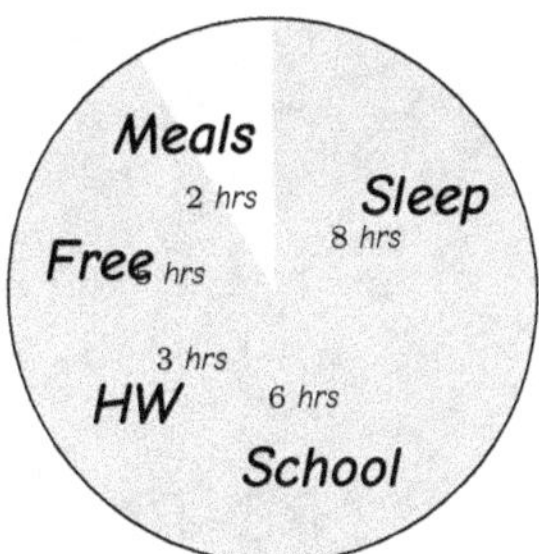

What percent of the day does the student spend on homework?

(A) 3%

(B) 8.3%

(C) 12.5%

(D) 25%

28. A teacher predicted that a spinner would land on red $\frac{1}{4}$ of the time. After 80 spins, red appeared 25 times. How do the observed and predicted results compare?

(A) They are very different — the model is wrong

(B) The observed ($\frac{25}{80} = 0.3125$) is close to the predicted (0.25)

(C) The observed result is exactly equal to the predicted result

(D) The predicted probability is higher than the observed

29. Three coins are flipped. What is the probability of getting NO heads (all tails)?

(A) $\frac{1}{2}$

(B) $\frac{1}{4}$

(C) $\frac{1}{8}$

(D) $\frac{3}{8}$

30. Describe how to use a standard die to simulate a $\frac{1}{3}$ probability event.

Your Answer:

 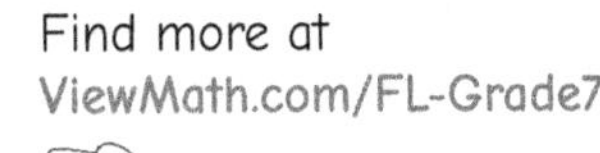
Get Online
Find more at
ViewMath.com/FL-Grade7

ViewMath.com

 # End of Practice Test 9 

Great job finishing the test!

 My Score

I got _____________ out of 30 questions right.

Check your answers in the Answer Key at the back of the book.

 *Review any questions you missed. That's how we learn!*

Check Your Score Online!

Visit **ViewMath Academy** to enter your answers and see which topics you need to review. You can also explore lessons, take quizzes, track your scores, and save your progress!

viewmath.com/score/7.1.FL.24

Or go to viewmath.com/score and enter code: 7.1.FL.24

Practice Test 10

30 Questions

✏ Before You Start ✏

- ✔ **Read each question carefully** before choosing your answer.
- ✔ **Show your work** on scratch paper when you need to.
- ✔ **Skip hard questions** and come back to them later.
- ✔ **Check your answers** when you're done.
- ✔ **Take your time** — there's no rush!

⭐ You've Got This! ⭐

Do your best and show what you know!

1. The graph below shows a proportional relationship. What is the constant of proportionality?

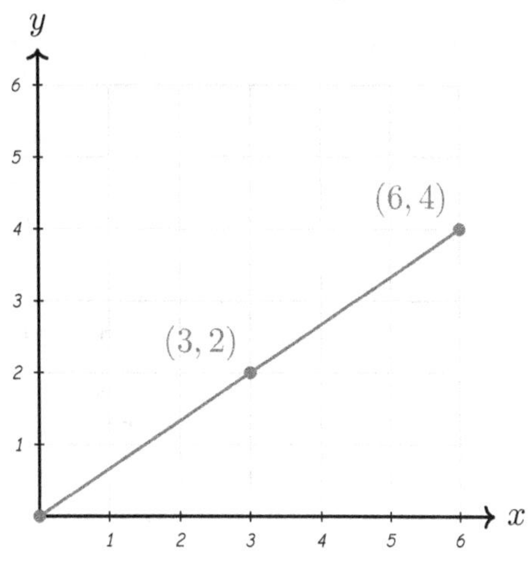

(A) $k = \frac{1}{2}$

(B) $k = \frac{2}{3}$

(C) $k = \frac{3}{2}$

(D) $k = 2$

2. The table shows the distance a bus travels over several hours. Using proportional reasoning, how far will the bus travel in 9 hours?

Hours	Miles
2	90
5	225
7	315
9	?

(A) 360 miles

(B) 385 miles

(C) 405 miles

(D) 450 miles

3. 18 is what percent of 72?

(A) 18%

(B) 20%

(C) 25%

(D) 36%

Get Online

Find more at
ViewMath.com/FL-Grade7

ViewMath.com

4. Which equation is equivalent to the proportion $\dfrac{n}{250} = \dfrac{16}{100}$?

(A) $n = 250 \div 16$

(B) $n = 16 \times 250$

(C) $100n = 250 \times 16$

(D) $16n = 250 \times 100$

5. A $1,600 loan at 5.5% simple interest accrues $264 interest. How long was the loan?

(A) 2 years

(B) 3 years

(C) 4 years

(D) 5 years

6. The bar graph shows the estimated and actual amounts of water (in liters) collected during an experiment over three days. On which day was the percent error the largest, and what was it?

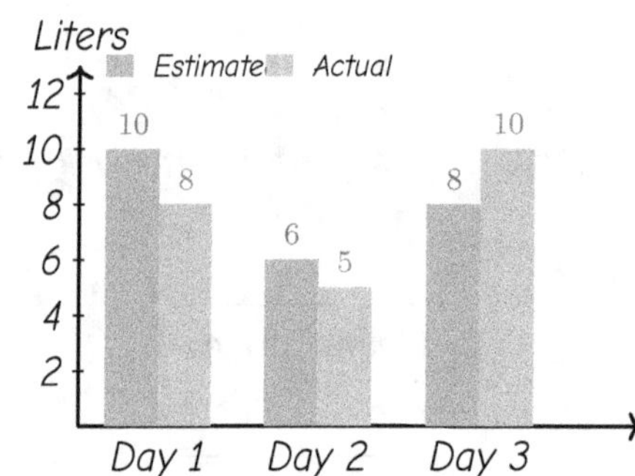

Your Answer:

7. What is $(-4) + (-4) + (-4)$?

Your Answer:

Get Online

Find more at
ViewMath.com/FL-Grade7

ViewMath.com

8. A bird is flying at 12 meters above the ground. A fish is swimming at -5 meters (below sea level). What is the distance between them?

Your Answer:

9. Which is the completely factored form of $8m + 20$?

(A) $2(4m + 10)$

(B) $4(2m + 5)$

(C) $8(m + 20)$

(D) $4(2m + 20)$

10. Simplify $(5n + 8) + (-2n + 3)$.

(A) $7n + 11$

(B) $3n + 5$

(C) $3n + 11$

(D) $7n + 5$

11. The total area of the figure below is 84 square cm. Find x.

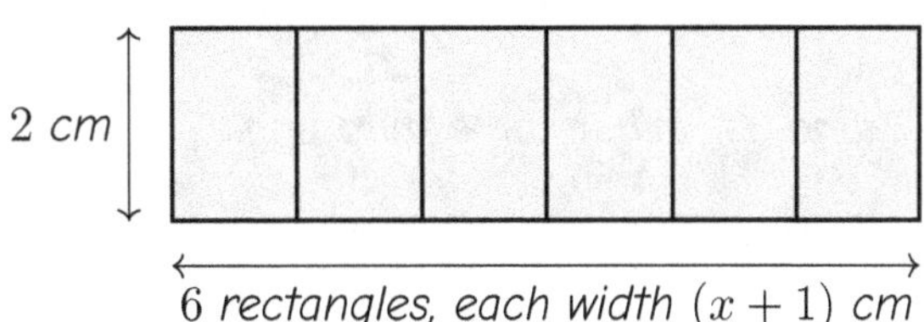

Your Answer:

12. Solve $0.6x - 2.4 = 1.2$.

(A) $x = 6$

(B) $x = -2$

(C) $x = 2$

(D) $x = 0.6$

Find more at
ViewMath.com/FL-Grade7

Get Online

ViewMath.com

13. *The balance scale tips to the left. Write and solve the inequality that describes this situation.*

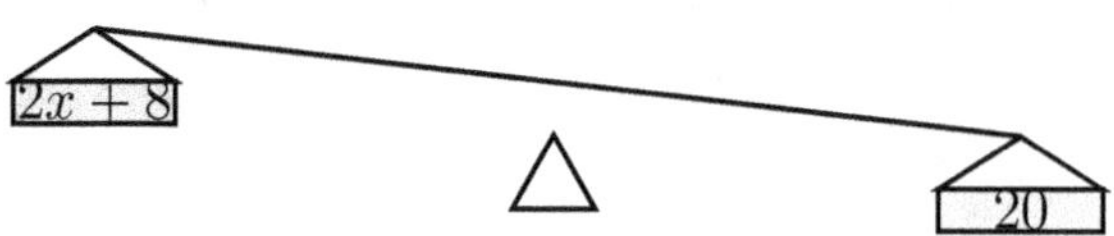

Your Answer

14. *Solve $6 - 2x \leq -4$ and describe the graph.*

(A) *Closed circle at 5, shade right*

(B) *Open circle at 5, shade right*

(C) *Closed circle at 5, shade left*

(D) *Closed circle at -5, shade right*

15. *A drawing uses 1 cm = 5 km. A road is 7 cm long. Another map shows the same road as 3.5 cm. What is the scale of the second map?*

(A) *1 cm = 2.5 km*

(B) *1 cm = 5 km*

(C) *1 cm = 10 km*

(D) *1 cm = 17.5 km*

16. *Two angles of a triangle are 72° and 53°. What is the third angle?*

Your Answer

17. *A hexagonal prism is sliced parallel to its base. What is the cross-section?*

(A) *Rectangle*

(B) *Triangle*

(C) *Hexagon*

(D) *Circle*

Find more at
ViewMath.com/FL-Grade7

ViewMath.com

18. Two lines intersect as shown. Find the values of a, b, and c.

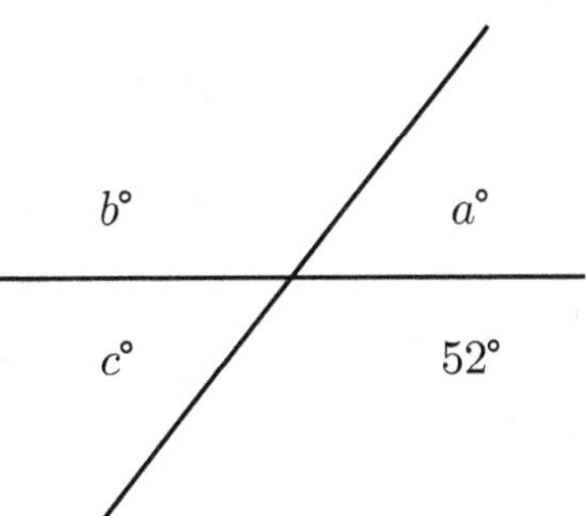

Your Answer:

19. If the radius of a circle is $\frac{3}{4}$ inch, what is the diameter?

(A) $\frac{3}{8}$ inch

(B) $\frac{3}{4}$ inch

(C) $1\frac{1}{2}$ inches

(D) 3 inches

20. A cube has a surface area of 150 cm². What is the side length of the cube?

(A) 5 cm

(B) 10 cm

(C) 25 cm

(D) 15 cm

21. What is the volume of a rectangular prism with length 6 cm, width 4 cm, and height 3 cm?

(A) $13\ cm^3$

(B) $24\ cm^3$

(C) $48\ cm^3$

(D) $72\ cm^3$

Find more at
ViewMath.com/FL-Grade7

ViewMath.com

22. If you double the radius of a cylinder but keep the height the same, what happens to the volume?

(A) It doubles

(B) It triples

(C) It quadruples

(D) It increases by a factor of 8

23. A radio station asks listeners to call in and vote for their favorite song. Explain why this is not a random sample.

Your Answer:

24. The histograms show the distribution of weekly study hours for two groups.

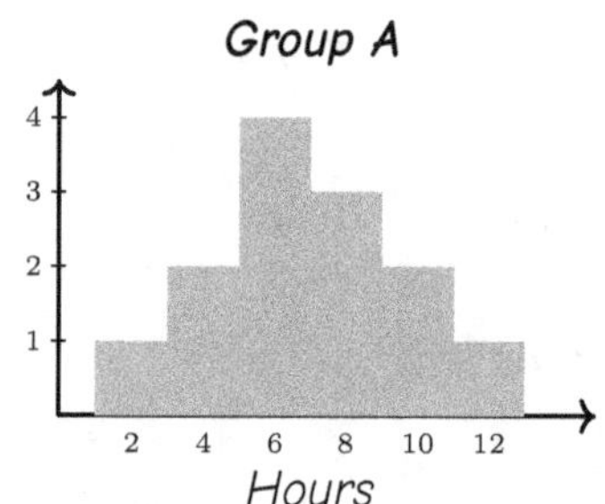

Group A

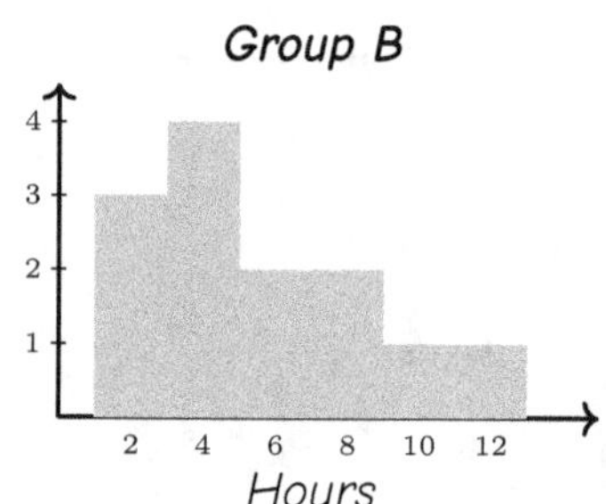

Group B

Which group tends to study more hours per week?

(A) Group A, because most of its data is at higher values

(B) Group B, because most of its data is at higher values

(C) Group A, because it has more spread

(D) They study the same amount

25. Group A: $\{60, 65, 70, 75, 80\}$. Group B: $\{40, 55, 70, 85, 100\}$. Both have a mean of 70. Which group is more variable?

(A) Group A

(B) Group B

(C) They are equally variable

(D) Cannot be determined

Find more at
ViewMath.com/FL-Grade7

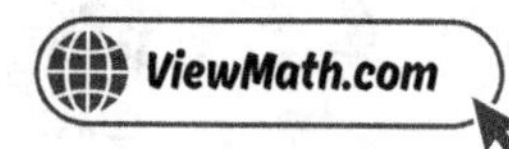
ViewMath.com

26. The data set is: $28, 31, 33, 36, 40, 42$. What is the median?

<blockquote>Your Answer:</blockquote>

27. The full circle in a circle graph represents:

(A) 50%   (B) 100% or 360°

(C) 180°   (D) The largest category

28. Look at the spinner below. Which probability model correctly represents this spinner?

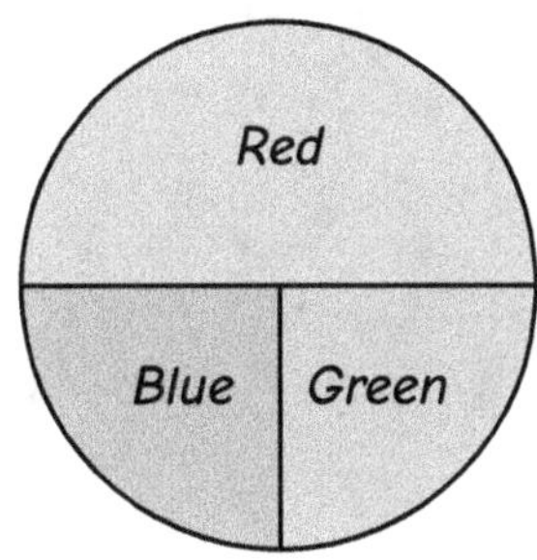

(A) $P(Red) = \frac{1}{3}$, $P(Blue) = \frac{1}{3}$, $P(Green) = \frac{1}{3}$ (B) $P(Red) = \frac{1}{2}$, $P(Blue) = \frac{1}{4}$, $P(Green) = \frac{1}{4}$

(C) $P(Red) = \frac{1}{4}$, $P(Blue) = \frac{1}{4}$, $P(Green) = \frac{1}{2}$ (D) $P(Red) = \frac{1}{2}$, $P(Blue) = \frac{1}{2}$, $P(Green) = 0$

29. Two dice are rolled. What is the probability that the product (not the sum) of the two numbers is 12?

(A) $\frac{2}{36}$ (B) $\frac{4}{36}$

(C) $\frac{3}{36}$ (D) $\frac{6}{36}$

Find more at
ViewMath.com/FL-Grade7

ViewMath.com

30. A student simulates rolling two dice 36 times to see how often the sum is 7. They get a sum of 7 a total of 8 times. How does this compare to the expected number?

(A) Expected: 6; the simulation result (8) is slightly higher

(B) Expected: 6; the simulation result (8) is much higher

(C) Expected: 12; the simulation result (8) is lower

(D) Expected: 7; the simulation result (8) is exact

Get Online

Find more at
ViewMath.com/FL-Grade7

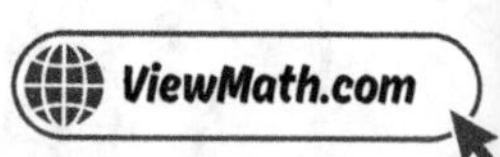

 ## ⭐ *End of Practice Test 10* ⭐ 

Great job finishing the test!

☑ *My Score*

I got _____________ out of 30 questions right.

*Check your answers in the **Answer Key** at the back of the book.*

💡 *Review any questions you missed. That's how we learn!*

📊 *Check Your Score Online!*

*Visit **ViewMath Academy** to enter your answers and see which topics you need to review. You can also explore lessons, take quizzes, track your scores, and save your progress!*

viewmath.com/score/7.1.FL.25

Or go to viewmath.com/score and enter code: 7.1.FL.25

Answer Key & Explanations

Answer Key

First try each test on your own, then check your work here.

✓ Practice Test 1 — Answer Key

 $k = 2.5$; missing $x = 6$  C  D  B  C  $\approx 4.2\%$  A  B

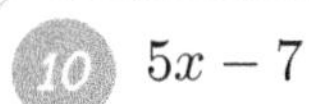 $-4(2x - 3y + 1)$  $5x - 7$ 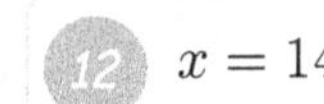 C 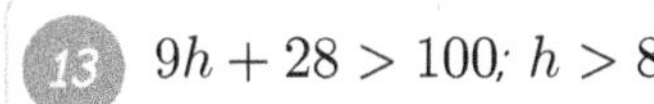 $x = 14$  $9h + 28 > 100$; $h > 8$  B

 5 cm  B  Any two of: square, rectangle, triangle, pentagon, hexagon

 $x = 6$; the angles are 43°  C  C  B  B  B  B  B

 C  C  0.4  B  48

💡 Time to Learn! 💡

Review the explanations below, **especially for the questions you missed**.

Understanding why each answer is correct builds stronger problem-solving skills.

Tip: Circle any questions you got wrong, then read their explanation carefully.

📖 Practice Test 1 — Detailed Explanations

1. $k = \frac{7.5}{3} = 2.5$. For the missing row: $15 = 2.5x$ so $x = 15 \div 2.5 = 6$.

Find more at
ViewMath.com/FL-Grade7

ViewMath.com

2 Scale: $2 \text{ cm} = 30 \text{ km}$, so $1 \text{ cm} = 15 \text{ km}$. For 6.5 cm: $15 \times 6.5 = 97.5 \text{ km}$.

3 The shaded portion is $3 \times 15 = 45$. The total is 75. Percent: $45 \div 75 = 0.60 = 60\%$.

4 $\frac{n}{500} = \frac{68}{100}$. Cross-multiply: $100n = 34{,}000$. Divide: $n = 340$.

5 Rates over 2 years: 500: $50/(500 \times 2) = 5\%$. $1{,}000$: $75/(1{,}000 \times 2) = 3.75\%$. 800: $100/(800 \times 2) = 6.25\%$. The $800 deposit had the highest rate.

6 $\frac{|5-4.8|}{4.8} \times 100 = \frac{0.2}{4.8} \times 100 \approx 4.17 \approx 4.2\%$.

7 $(-4) + (-3)$ means start at -4 and move 3 units left (adding a negative). Model A shows the arrow going from -4 to -7, which is correct.

8 $0 - (-6) = 0 + 6 = 6$. Subtracting a negative is the same as adding.

9 GCF of 8, 12, and 4 is 4. Factor out -4: $-8x \div (-4) = 2x$, $12y \div (-4) = -3y$, $-4 \div (-4) = 1$. Result: $-4(2x - 3y + 1)$.

10 Combine: $(-3x + 8x) + (5 - 12) = 5x - 7$.

11 Distance from A to B is 15. $3(x + 2) = 15$. Divide by 3: $x + 2 = 5$. Subtract 2: $x = 3$.

12 Subtract 1.5: $\frac{x}{4} = 3.5$. Multiply by 4: $x = 14$. The point is at 14 on the number line.

13 $9h + 28 > 100$. Subtract 28: $9h > 72$. Divide by 9: $h > 8$. You need to work more than 8 hours.

14 $x > 3$ uses an open circle because 3 is not included (strictly greater than).

Find more at
ViewMath.com/FL-Grade7

ViewMath.com

15. The scale ratio is $\frac{7}{7} = 1$. The length stays the same: $5 \times 1 = 5$ cm.

16. Two sides and the angle between them is the SAS (Side-Angle-Side) condition.

17. A cube can be sliced to produce squares, rectangles, triangles, pentagons, or hexagons depending on the angle and position of the cut.

18. Vertical angles are equal: $7x + 1 = 5x + 13$. Subtract $5x$: $2x + 1 = 13$. Subtract 1: $2x = 12$. Divide: $x = 6$. Angle: $7(6) + 1 = 43°$.

19. The path through the center is a diameter: $d = 2r = 2 \times 8 = 16$ feet.

20. $94 = 2(5 \times 3) + 2(5 \times h) + 2(3 \times h) = 30 + 10h + 6h = 30 + 16h$. So $16h = 64$, $h = 4$ cm.

21. $s^3 = 64$. $s = 4$ cm because $4^3 = 64$.

22. $V = \pi r^2 h$. $314 = 3.14 \times 25 \times h = 78.5h$. $h = 314 \div 78.5 = 4$ cm.

23. Only surveying recent buyers introduces bias — they may feel differently than all customers.

24. The means differ (45 vs. 52) but both have the same range (12), so they have different centers and similar spread.

25. $|4 - 8| + |6 - 8| + |8 - 8| + |10 - 8| + |12 - 8| = 4 + 2 + 0 + 2 + 4 = 12$. MAD $= \frac{12}{5} = 2.4$.

26. Values greater than 25: $27, 31, 35, 39, 42$ — that's 5 values.

27. $\frac{16}{40} = 0.40 = 40\%$.

Find more at
ViewMath.com/FL-Grade7

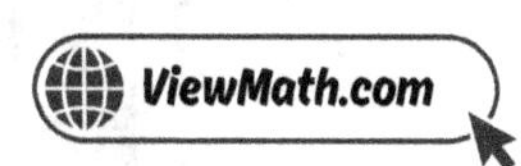

28 $P(Z) = 1 - 0.15 - 0.45 = 0.40.$

29 $P(\text{spinner} = 3) = \frac{1}{4}$, $P(\text{die} = 3) = \frac{1}{6}$. $P = \frac{1}{4} \times \frac{1}{6} = \frac{1}{24}$.

30 $80 \times 0.60 = 48.$

☑ Practice Test 2 — Answer Key

| 1 | B | 2 | 21 feet | 3 | 90 students | 4 | A | 5 | A | 6 | C | 7 | C | 8 | −6 | 9 | B |

10 $2a + 8b$ 11 B   12 $x = 8$ 13 $9s + 35 \leq 200$; $s \leq 18$ 14 $x \geq -6$ 15 B   16 B

17 A   18 B   19 2.8 inches   20 C   21 B   22 748 cm^2 23 C

24 0.5 MADs; no, the difference is not meaningful   25 A   26 Range = 67; Median = 33.5   27 B

28 B   29 C   30 B

💡 Time to Learn! 💡

Review the explanations below, **especially for the questions you missed**.

Understanding why each answer is correct builds stronger problem-solving skills.

Tip: Circle any questions you got wrong, then read their explanation carefully.

📖 Practice Test 2 — Detailed Explanations

Find more at
ViewMath.com/FL-Grade7

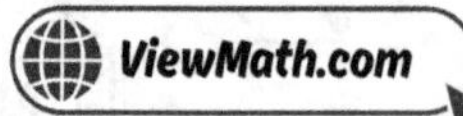

1. $k = \frac{54}{2} = 27$, $\frac{135}{5} = 27$, $\frac{216}{8} = 27$. The car gets 27 miles per gallon.

2. Scale: 1 inch $= 6$ feet. For 3.5 inches: $6 \times 3.5 = 21$ feet.

3. 45% of 200: $0.45 \times 200 = 90$ students take the bus.

4. Part (42) over whole (w) equals percent (60) over 100: $\frac{42}{w} = \frac{60}{100}$.

5. 6 months $= 0.5$ years. $I = 900 \times 0.08 \times 0.5 = \$36$.

6. A: $\frac{|180-200|}{200} = 10\%$. B: $\frac{|220-200|}{200} = 10\%$. Both have 10% error.

7. Different signs: subtract $12 - 7 = 5$. The larger absolute value is 12 (negative), so the answer is -5.

8. $(-6) - (-6) = (-6) + 6 = 0$. Then $0 - 6 = -6$.

9. GCF is 3. Factor out -3: $-9x \div (-3) = 3x$ and $-6 \div (-3) = 2$. Result: $-3(3x + 2)$. Check: $-3(3x + 2) = -9x - 6$ ✓.

10. Distribute: $4a + 3b - 2a + 5b$. Combine: $(4a - 2a) + (3b + 5b) = 2a + 8b$.

11. Each pays $\frac{c}{3} + 4 = 16$. This means $\frac{c}{3} = 12$, so $c = 36$. Or: total paid $= 3 \times 16 = 48$, wrapping $= 3 \times 4 = 12$, gift $= 48 - 12 = 36$.

12. Subtract 4.2: $-0.3x = -2.4$. Divide by -0.3: $x = 8$. Check: $-0.3(8) + 4.2 = -2.4 + 4.2 = 1.8$ ✓.

13. Total cost: $9s + 35 \le 200$. Subtract 35: $9s \le 165$. Divide by 9: $s \le 18.\overline{3}$. Since s must be a whole number, $s \le 18$.

Find more at
ViewMath.com/FL-Grade7

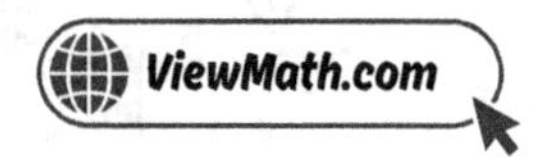

14 Closed circle means -6 is included, shading right means values greater than or equal to -6: $x \geq -6$.

15 A smaller scale means each cm represents more real distance, so fewer cm are needed. The drawing gets smaller. The shape stays the same.

16 $180° - 35° - 75° = 70°$.

17 A vertical cut through a cube perpendicular to the base typically produces a rectangle. If the cut goes through the center parallel to a face, it can be a square.

18 Let the smaller be x. Then $x + (x + 25) = 180$. Combine: $2x + 25 = 180$. Solve: $2x = 155$, $x = 77.5°$. The other: $77.5 + 25 = 102.5°$.

19 $r = d \div 2 = 5.6 \div 2 = 2.8$ inches.

20 A rectangular prism has 6 faces: top, bottom, front, back, left, and right.

21 Space volume: $6 \times 4 \times 3 = 72$ ft^3. Box volume: $2 \times 2 \times 3 = 12$ ft^3. Number: $72 \div 12 = 6$ boxes.

22 $SA = 2\pi r^2 + 2\pi rh = 2 \times \frac{22}{7} \times 49 + 2 \times \frac{22}{7} \times 7 \times 10 = 308 + 440 = 748$ cm^2.

23 In a random sample, every member of the population has an equal chance of being chosen.

24 $\frac{54-50}{8} = 0.5$ MADs. Since $0.5 < 2$, the difference is small relative to the variability and not considered meaningful.

25 Class M: $Q_3 - Q_1 = 17 - 12 = 5$. Class N: $18 - 13 = 5$. Both have IQR $= 5$.

26 Data (20 values): $5, 8, 9, 12, 14, 16, 20, 25, 28, 32, 35, 37, 39, 41, 45, 50, 58, 63, 67, 72$. Range $= 72 - 5 = 67$. Median $= \frac{32+35}{2} = 33.5$.

Find more at
ViewMath.com/FL-Grade7

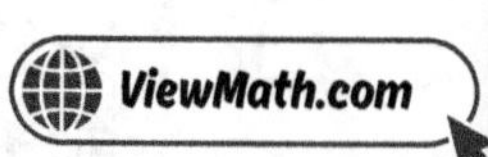

27 A circle graph (pie chart) shows how a whole quantity is divided into categories.

28 Each color should appear about 50 times (200×0.25). The results $(55, 48, 52, 45)$ are all close to 50, supporting the model.

29 Outcomes with at least one tail: HT, TH, TT — that is 3 out of 4. $P = \frac{3}{4}$.

30 With 4 equal sections, each should appear about 25 times in 100 spins. The results $(26, 24, 28, 22)$ are all close to 25, suggesting the spinner is fair.

✔ Practice Test 3 — Answer Key

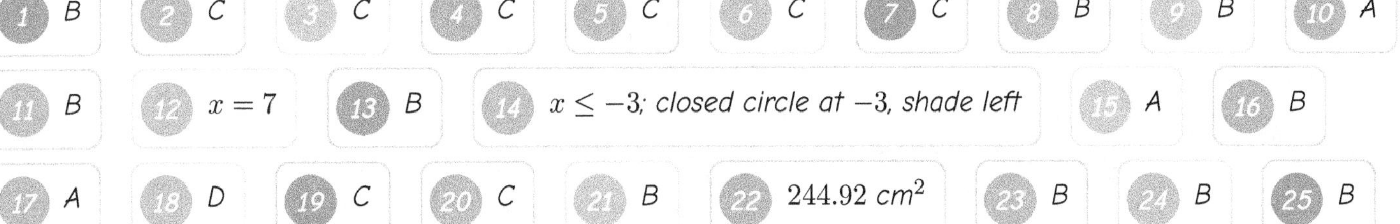

1 B	2 C	3 C	4 C	5 C	6 C	7 C	8 B	9 B	10 A
11 B	12 $x = 7$	13 B	14 $x \leq -3$; closed circle at -3, shade left		15 A	16 B
17 A	18 D	19 C	20 C	21 B	22 $244.92\ cm^2$	23 B	24 B	25 B
26 11	27 A	28 B	29 C	30 A

💡 Time to Learn! 💡

Review the explanations below, **especially for the questions you missed**.

Understanding why each answer is correct builds stronger problem-solving skills.

Tip: Circle any questions you got wrong, then read their explanation carefully.

📖 Practice Test 3 — Detailed Explanations

Find more at
ViewMath.com/FL-Grade7

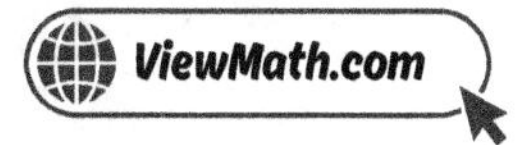

1. If $k = \frac{3}{4}$, then $y = \frac{3}{4}x$. For $x = 8$: $y = \frac{3}{4} \times 8 = 6$. So $(8, 6)$ works.

2. Per person: $\frac{3}{2} \div 4 = \frac{3}{8}$ cup. For 10: $\frac{3}{8} \times 10 = \frac{30}{8} = 3\frac{3}{4}$ cups.

3. Divide: $63 \div 0.90 = 70$.

4. Proportion: $\frac{n}{75} = \frac{40}{100}$ gives $n = 30$. Equation: $n = 0.40 \times 75 = 30$. Both give 30.

5. $I = 1{,}200 \times 0.035 \times 2 = \$84$

6. $\frac{|4.5 - 5|}{5} \times 100 = \frac{0.5}{5} \times 100 = 10\%$.

7. $(-50) + 75 = 25$. The deposit is larger than the overdraft, so the balance becomes positive.

8. Subtraction means adding the opposite: $(-5) - 3 = (-5) + (-3)$.

9. GCF of 16, 24, and 8 is 8. Divide each: $16x \div 8 = 2x$, $24y \div 8 = 3y$, $-8 \div 8 = -1$. Result: $8(2x + 3y - 1)$.

10. Distribute: $3.5n - 2.1 - 1.5n - 0.9$. Combine: $(3.5n - 1.5n) + (-2.1 - 0.9) = 2n - 3$.

11. The student wrote $3x + 4$ instead of $3x + 12$. Distributing correctly gives $3x + 12 = 30$.

12. Add 7.5: $2.5x = 17.5$. Divide by 2.5: $x = 7$. Check: $2.5(7) - 7.5 = 17.5 - 7.5 = 10$ ✓.

13. Divide by -4 and flip the sign: $x < -5$.

14. Subtract 15: $-5x \geq 15$. Divide by -5 and flip: $x \leq -3$. Closed circle at -3, shade left.

Find more at
ViewMath.com/FL-Grade7

15. Scale ratio: $\frac{10}{40} = \frac{1}{4}$. Every old length is multiplied by $\frac{1}{4}$ on the new drawing.

16. Third angle: $180° - 25° - 25° = 130°$. Two equal angles means two equal sides, making it isosceles.

17. Cut A (horizontal, parallel to the base) produces a circle. Cut B (vertical through the apex) produces a triangle.

18. Let the supplement be x. The angle is $4x$. Sum: $x + 4x = 180$. So $5x = 180$ and $x = 36°$. The angle: $4(36) = 144°$.

19. A vertex is a corner point found on polygons, not circles. Center, radius, and diameter are all parts of a circle.

20. $SA = 2(30 \times 20) + 2(30 \times 6) + 2(20 \times 6) = 1200 + 360 + 240 = 1,800 \ cm^2$.

21. $B = 8 \times 6 = 48 \ cm^2$. $h = V \div B = 240 \div 48 = 5 \ cm$.

22. $r = 3 \ cm$. $SA = 2\pi r^2 + 2\pi rh = 2(3.14)(9) + 2(3.14)(3)(10) = 56.52 + 188.4 = 244.92 \ cm^2$.

23. A random sample provides an estimate. About 35% of the population likely prefers pizza, but the exact percentage may differ slightly.

24. Dot plots show every data point, which is useful for small data sets where individual values matter.

25. A wider range indicates more variability, not better performance. Class A has a higher median and is more consistent (smaller IQR and range).

26. Count leaves: $3 + 5 + 2 + 1 = 11$.

27. Class A: 45% of 20 = 9. Class B: 30% of 30 = 9. Same number: 9 each.

Find more at
ViewMath.com/FL-Grade7

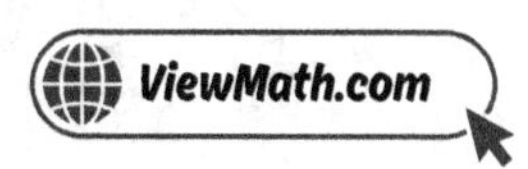

28 The probability of drawing red ($\frac{4}{5}$) is different from drawing blue ($\frac{1}{5}$), so this is a non-uniform model.

29 The only way to get a sum of 2 is $(1,1)$. $P = \frac{1}{36}$.

30 $P = \frac{6}{40} = 0.15$. The theoretical probability is $\frac{1}{8} = 0.125$, close to the simulation result.

✓ Practice Test 4 — Answer Key

1 B	2 9 trips	3 D	4 36	5 C	6 C	7 C	8 8 units	9 D
10 B	11 C	12 $x = 12$	13 B	14 $x > 2$; open circle at 2, shade right	15 C	16 C
17 C	18 A	19 B	20 C	21 C	22 C	23 B
24 Yes; the means are about 2.5 to 3.3 MADs apart	25 A	26 A	27 B	28 C	29 $\frac{3}{8}$
30 B

💡 Time to Learn! 💡

Review the explanations below, **especially for the questions you missed.**

Understanding why each answer is correct builds stronger problem-solving skills.

Tip: Circle any questions you got wrong, then read their explanation carefully.

📖 Practice Test 4 — Detailed Explanations

1. $k = \frac{y}{x} = \frac{12}{3} = 4$. Check: $\frac{20}{5} = 4$ and $\frac{32}{8} = 4$.

2. Per trip: $540 \div 3 = 180$ packages. Trips needed: $1{,}620 \div 180 = 9$.

3. Divide the part by the percent: $9 \div 0.12 = 75$.

4. $\frac{n}{80} = \frac{45}{100}$. Cross-multiply: $100n = 3600$. Divide: $n = 36$ tomato plants.

5. $I = Prt = 500 \times 0.06 \times 3 = \$90$.

6. Absolute value ensures the percent error is positive whether the estimate was too high or too low.

7. $(-9) + n = -2$ means $n = -2 - (-9) = -2 + 9 = 7$.

8. Distance $= |-11 - (-3)| = |-11 + 3| = |-8| = 8$ units.

9. $4(-3y + 2) = -12y + 8$ ✓. Also $-4(3y - 2) = -12y + 8$ ✓. Both are valid factored forms.

10. Distribute the negative: $6a - 2 - 3a - 5$. Combine: $(6a - 3a) + (-2 - 5) = 3a - 7$.

11. Divide by 7: $x - 2 = 0$. Add 2: $x = 2$. Check: $7(2 - 2) = 7(0) = 0$ ✓.

12. Multiply every term by 4: $2x + x = 36$. Combine: $3x = 36$. Divide by 3: $x = 12$. Check: $\frac{12}{2} + \frac{12}{4} = 6 + 3 = 9$ ✓.

13. The graph shows $x \leq -2$ (closed circle, shading left). $-4x \geq 8$: divide by -4 and flip: $x \leq -2$ ✓.

14. Subtract 10: $-4x < -8$. Divide by -4 and flip: $x > 2$. Open circle at 2, shade right.

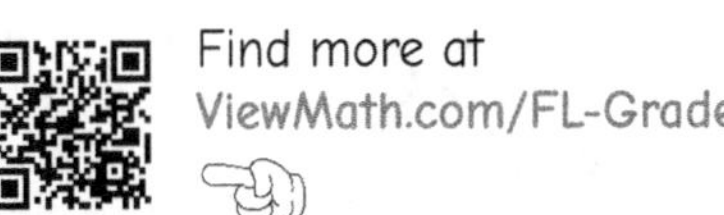

15 Scale ratio: $\frac{20}{10} = 2$. Every length on the new drawing is 2 times the old one. The drawing gets bigger because each cm now represents fewer real km.

16 Both triangles have the same angles but different side lengths. AAA determines shape but not size, so many triangles are possible.

17 A rectangular prism has flat faces and straight edges. Its cross-sections can be triangles, rectangles, pentagons, or hexagons, but never circles (since it has no curved surfaces).

18 Vertical angles are equal (130° and 130°). Adjacent angles are supplementary: $180° - 130° = 50°$. The four angles are 130°, 50°, 130°, 50°.

19 A diameter is made up of two radii placed end to end through the center.

20 The three pairs of faces have areas $7 \times 4 = 28$, $7 \times 3 = 21$, and $4 \times 3 = 12$ m^2. The largest is 28 m^2.

21 $V = 50 \times 30 \times 40 = 60{,}000$ cm^3.

22 $SA = 2\pi r^2 + 2\pi rh = 2(3.14)(9) + 2(3.14)(3)(7) = 56.52 + 131.88 = 188.4$ cm^2.

23 Selecting every 10th student from the school roster is systematic and gives all students a chance, making it the least biased.

24 Difference $= 10$. Using MAD of 4: $\frac{10}{4} = 2.5$. Using MAD of 3: $\frac{10}{3} \approx 3.3$. Both are greater than 2, so the difference is meaningful.

25 Difference $= 60 - 50 = 10$. In MADs: $\frac{10}{3} \approx 3.3$.

26 Stems are tens digits $(4, 5, 6)$, leaves are ones digits listed in order. Only choice A has correct stems and ordered leaves.

27 $100 + 80 + 60 + 100 = 340°$, which is $20°$ short of $360°$. The graph is incorrect.

28 $P(tails) = 1 - 0.60 = 0.40$. Since heads and tails have different probabilities, this is a non-uniform model.

29 Outcomes with exactly 2 tails: HTT, THT, TTH — that is 3 out of 8. $P = \frac{3}{8}$.

30 Each coin flip has a 50% chance of heads (pass). Flipping 3 coins and counting exactly 2 heads simulates exactly 2 passes.

✅ Practice Test 5 — Answer Key

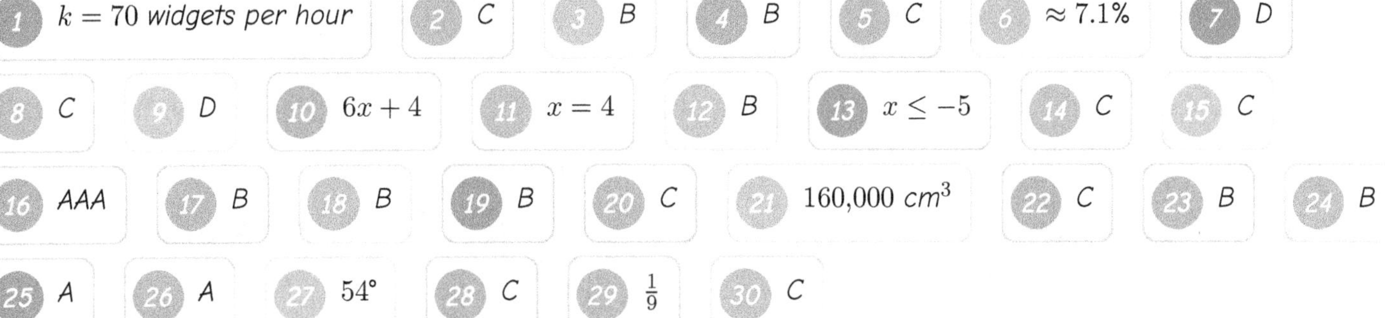

1	$k = 70$ widgets per hour	2 C	3 B	4 B	5 C	6 $\approx 7.1\%$	7 D
8 C	9 D	10 $6x + 4$	11 $x = 4$	12 B	13 $x \leq -5$	14 C	15 C
16 AAA	17 B	18 B	19 B	20 C	21 $160{,}000\ cm^3$	22 C	23 B	24 B
25 A	26 A	27 $54°$	28 C	29 $\frac{1}{9}$	30 C

💡 Time to Learn! 💡

Review the explanations below, **especially for the questions you missed**.

Understanding why each answer is correct builds stronger problem-solving skills.

Tip: Circle any questions you got wrong, then read their explanation carefully.

📖 Practice Test 5 — Detailed Explanations

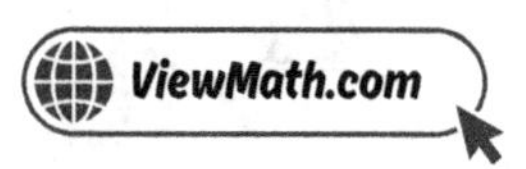

1. $k = \frac{175}{2.5} = 70$ widgets per hour.

2. Set up a proportion: $\frac{2}{15} = \frac{7}{d}$. Cross-multiply: $2d = 105$, so $d = 52.5$ miles.

3. Divide: $54 \div 200 = 0.27 = 27\%$.

4. $\frac{n}{80} = \frac{37.5}{100}$. Cross-multiply: $100n = 3000$. Divide: $n = 30$.

5. $300 = 2{,}000 \times 0.03 \times t$ ⊠ $300 = 60t$ ⊠ $t = 5$ years.

6. $\frac{|45-42|}{42} \times 100 = \frac{3}{42} \times 100 \approx 7.14 \approx 7.1\%$.

7. A number plus its opposite (additive inverse) always equals 0.

8. $4 - (-11) = 4 + 11 = 15°F$.

9. $5(-3a + 5) = -15a + 25$ ✓. $-5(3a - 5) = -15a + 25$ ✓. Both are correct factored forms.

10. Total before coupon: $3x + 7 + 5x - 2 = 8x + 5$. After coupon: $(8x + 5) - (2x + 1) = 8x + 5 - 2x - 1 = 6x + 4$

11. Divide by 4: $x + 9 = 13$. Subtract 9: $x = 4$. Check: $4(4 + 9) = 4(13) = 52$ ✓.

12. Multiplying every term by 4 (the LCD) clears the fractions: $x + 2 = 12$.

13. Subtract 8: $-3x \geq 15$. Divide by -3 and flip: $x \leq -5$.

14. $x \geq -4$ includes -4 and everything to the right. $-5 < -4$, so -5 is not a solution.

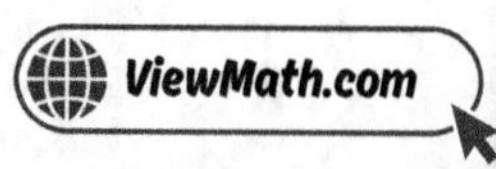

15 Scale ratio: $\frac{5}{2} = 2.5$. New dimensions: $3 \times 2.5 = 7.5$ cm and $4 \times 2.5 = 10$ cm. New area: $7.5 \times 10 = 75$ cm^2.

16 Three angles determine the shape but not the size. Infinitely many similar triangles can share the same three angle measures.

17 The shape of a cross-section depends on where you cut and at what angle. Different cuts through the same figure can produce different shapes.

18 Vertical angles are always equal. If one is 72°, the vertical angle is also 72°.

19 In standard circle diagrams, O typically labels the center of the circle.

20 $SA = 2(6 \times 6) + 2(6 \times 10) + 2(6 \times 10) = 72 + 120 + 120 = 312$ ft^2.

21 $V = 80 \times 40 \times 50 = 160{,}000$ cm^3.

22 $SA = 2\pi r^2 + 2\pi rh = 2(3.14)(25) + 2(3.14)(5)(8) = 157 + 251.2 = 408.2$ cm^2.

23 The sample size is the number of individuals selected — 500 patients.

24 The mean tells us the average performance. Class B scored higher on average, but individual scores may overlap.

25 Difference $= 82 - 76 = 6$. In MADs: $\frac{6}{5} = 1.2$. Since $1.2 < 2$, the difference is not considered very meaningful.

26 The values are $22, 22, 25, 28$. The value 22 appears twice (most often), so it is the mode.

27 $0.15 \times 360° = 54°$.

Find more at
ViewMath.com/FL-Grade7

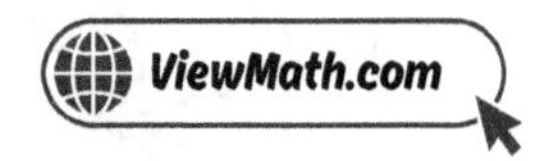

28　There are 2 B tiles out of 5 total. $P(B) = \frac{2}{5}$.

29　Products of 6: $(1,6), (2,3), (3,2), (6,1)$ — 4 outcomes. $P = \frac{4}{36} = \frac{1}{9}$.

30　Before choosing a model or running trials, you must first identify what event you're simulating and what the possible outcomes are.

✅ Practice Test 6 — Answer Key

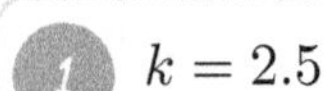 $k = 2.5$　 C　 C　 C　 C　 C　 C　 C　 A

 A　 A　 A　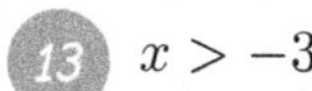 $x > -3$　 A　 C　 C　 6 cm by 4 cm

 40　 C　 96 m^2　 D　 326.56 cm^2　 C　 B　 B　 C

 D　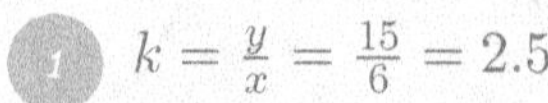 B　 C　 B

💡 Time to Learn! 💡

Review the explanations below, **especially for the questions you missed**.

Understanding why each answer is correct builds stronger problem-solving skills.

Tip: Circle any questions you got wrong, then read their explanation carefully.

📖 Practice Test 6 — Detailed Explanations

1　$k = \frac{y}{x} = \frac{15}{6} = 2.5$.

2 Rate: $120 \div 8 = 15$ pages/min. Time for 450: $450 \div 15 = 30$ min.

3 $35\% = 0.35$. Multiply: $0.35 \times 480 = 168$ fiction books.

4 $\frac{24}{80} = \frac{p}{100}$. Cross-multiply: $80p = 2400$. Divide: $p = 30\%$.

5 A: $1{,}000 \times 0.05 \times 2 = \$100$. B: $1{,}000 \times 0.02 \times 5 = \$100$. Both earn $\$100$.

6 $\frac{|30-28|}{28} \times 100 = \frac{2}{28} \times 100 \approx 7.1\%$.

7 $(-4) + 9 = 5$. The positive number (9) has a larger absolute value than 4.

8 $a - b = (-8) - 5 = (-8) + (-5) = -13$.

9 GCF of 15 and 10 is 5. Factor: $15a \div 5 = 3a$ and $-10 \div 5 = -2$. Result: $5(3a - 2)$.

10 Distribute: $2y + 7 - 5y + 3$. Combine: $(2y - 5y) + (7 + 3) = -3y + 10$.

11 Divide by 10: $x - 5 = 3$. Add 5: $x = 8$. Check: $10(8 - 5) = 10(3) = 30$ ✓.

12 From $x = 2$ to $x = 4$, y increases by 1 while x increases by 2, so rate $= 0.5$. When $x = 2$: $0.5(2) + 1.5 = 2.5$ ✓.

13 Add 4: $-7x < 21$. Divide by -7 and flip: $x > -3$.

14 You can spend at most $\$30$: $5h \le 30$, so $h \le 6$. Closed circle at 6, shade left (and $h \ge 0$).

15 Scale ratio: $\frac{12}{4} = 3$. New width: $0.5 \times 3 = 1.5$ in.

Find more at
ViewMath.com/FL-Grade7

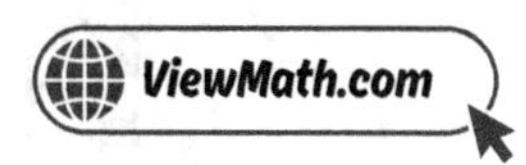

16 Although $90 + 100 + (-10) = 180$, every angle in a triangle must be greater than 0°. A negative angle is impossible.

17 A cut parallel to the base of a rectangular prism produces a rectangle identical to the base: 6 cm by 4 cm.

18 Angles on a line sum to 180°: $50 + 2x + 10 + x = 180$. Combine: $3x + 60 = 180$. Solve: $3x = 120$, $x = 40$.

19 The diameter is always twice the radius: $d = 2r$.

20 Prism: $SA = 2(4 \times 4) + 2(4 \times 10) + 2(4 \times 10) = 32 + 80 + 80 = 192$ m^2. Cube: $6 \times 16 = 96$ m^2. Difference: $192 - 96 = 96$ m^2.

21 First: $8 \times 5 \times 4 = 160$ in^3. Second: $10 \times 4 \times 4 = 160$ in^3. They have the same volume.

22 $SA = 2\pi r^2 + 2\pi r h = 2(3.14)(16) + 2(3.14)(4)(9) = 100.48 + 226.08 = 326.56$ cm^2.

23 When people choose whether to participate, it is a voluntary response sample, which tends to be biased.

24 Histograms show the shape and distribution of large data sets, making it easy to compare two groups.

25 Same mean means same center. Different MADs mean one group's data is more spread out than the other's.

26 A stem-and-leaf plot shows every individual data value AND the overall shape/distribution of the data.

27 55% of $500 = 0.55 \times 500 = 275$.

28 The probabilities are $0.4, 0.3, 0.2, 0.1$ — all different, so this is non-uniform. They sum to 1.0, making it valid.

29 $P(\text{tails}) = \frac{1}{2}$ and $P(\text{even}) = \frac{3}{6} = \frac{1}{2}$. $P = \frac{1}{2} \times \frac{1}{2} = \frac{1}{4}$. Or: 3 favorable out of 12 total outcomes.

30 A simulation uses random numbers, coins, dice, or spinners to model a real-world event and estimate probabilities.

✅ Practice Test 7 — Answer Key

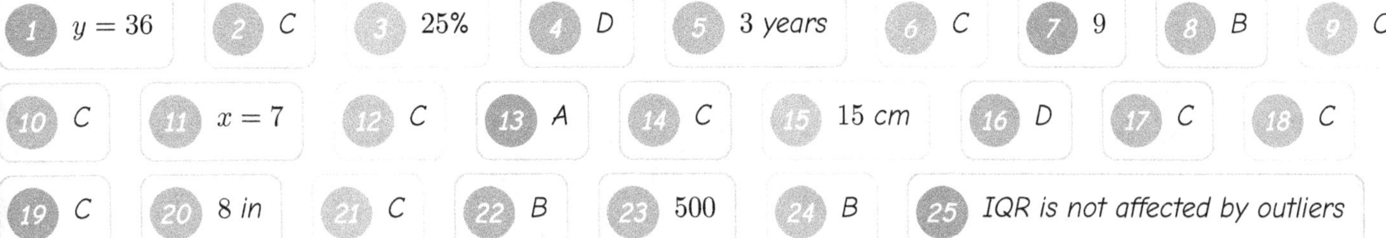

1 $y = 36$	**2** C	**3** 25%	**4** D	**5** 3 years	**6** C	**7** 9	**8** B	**9** C
10 C	**11** $x = 7$	**12** C	**13** A	**14** C	**15** 15 cm	**16** D	**17** C	**18** C
19 C	**20** 8 in	**21** C	**22** B	**23** 500	**24** B	**25** IQR is not affected by outliers		
26 B	**27** Sports: 40%, 144°; Music: 25%, 90°; Art: 15%, 54°; Reading: 20%, 72°					**28** B	**29** B	
30 C								

💡 Time to Learn! 💡

Review the explanations below, **especially for the questions you missed**.

Understanding why each answer is correct builds stronger problem-solving skills.

Tip: Circle any questions you got wrong, then read their explanation carefully.

📖 Practice Test 7 — Detailed Explanations

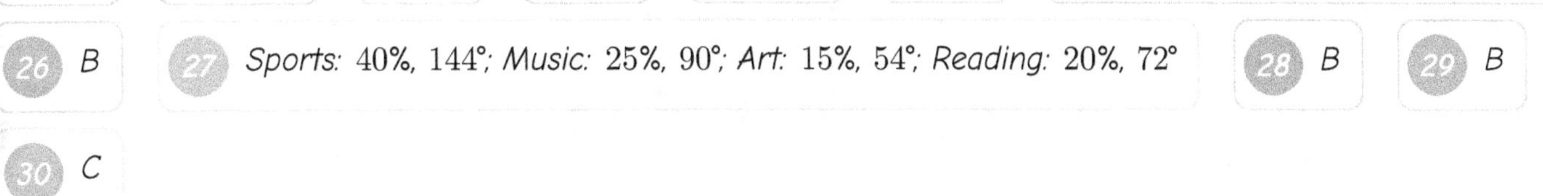

1 $k = \frac{20}{5} = 4$. So $y = 4 \times 9 = 36$.

Find more at
ViewMath.com/FL-Grade7

ViewMath.com

2   Rate: $350 \div 5 = 70$ bottles/hr. In 12 hours: $70 \times 12 = 840$ bottles.

3   Divide the part by the whole: $42 \div 168 = 0.25 = 25\%$.

4 $\frac{135}{w} = \frac{54}{100}$. Cross-multiply: $54w = 13{,}500$. Divide: $w = 250$.

5 $450 = 5{,}000 \times 0.03 \times t \Rightarrow 450 = 150t \Rightarrow t = 3$ years.

6 $\frac{|50-40|}{40} \times 100 = \frac{10}{40} \times 100 = 25\%$.

7 $(-15) + n = -6$, so $n = -6 - (-15) = -6 + 15 = 9$.

8   A: $5 - 12 = -7$. B: $(-3) + 10 = 7$. C: $(-8) + (-1) = -9$. D: $2 - 7 = -5$. The greatest is 7.

9   Factors of 12: $1, 2, 3, 4, 6, 12$. Factors of 18: $1, 2, 3, 6, 9, 18$. The greatest common factor is 6.

10   Any expression minus itself equals 0: $4x + 9 - 4x - 9 = 0$.

11 $3(2x + 5) = 57$. Divide by 3: $2x + 5 = 19$. Subtract 5: $2x = 14$. Divide by 2: $x = 7$.

12   Subtract 4.5: $-1.5x = -7.5$. Divide by -1.5: $x = 5$. Check: $-1.5(5) + 4.5 = -7.5 + 4.5 = -3$ ✓.

13   Subtract 4: $\frac{x}{3} < 6$. Multiply by 3: $x < 18$.

14   Open circle at 0 (not included) with shading left means $x < 0$, which is all numbers less than 0.

15   Scale ratio: $\frac{12}{4} = 3$. New length: $5 \times 3 = 15$ cm.

Find more at
ViewMath.com/FL-Grade7

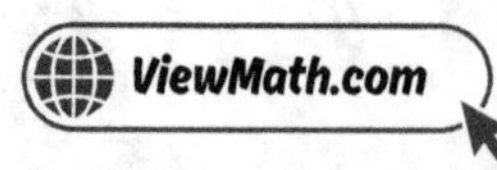

16 AAA determines the shape but not the size. Infinitely many similar triangles of different sizes share these angle measures.

17 A horizontal cut parallel to the base of a rectangular prism produces a rectangle the same shape as the base.

18 If two equal angles sum to $180°$, then each is $180 \div 2 = 90°$.

19 $r = d \div 2 = 26 \div 2 = 13$ inches.

20 $s^2 = 384 \div 6 = 64.$ So $s = 8$ in.

21 $B = \frac{1}{2} \times 5 \times 12 = 30$ cm^2. $V = 30 \times 15 = 450$ cm^3.

22 $V = \pi r^2 h = 3.14 \times 4 \times 5 = 62.8$ m^3.

23 $\frac{20}{80} = 25\%$. Then 25% of $2,000 = 500.$

24 Class 1 center ≈ 10. Class 2 center ≈ 13. Difference ≈ 3 push-ups. Class 2 is higher.

25 Range uses only the extreme values, which can be misleading with outliers. IQR measures the middle 50% and gives a more stable picture of typical variability.

26 The key tells readers what the stems and leaves represent, so they can correctly read the data values.

27 Sports: $\frac{48}{120} = 40\%$, $0.40 \times 360° = 144°$. Music: $\frac{30}{120} = 25\%$, $90°$. Art: $\frac{18}{120} = 15\%$, $54°$. Reading: $\frac{24}{120} = 20\%$, $72°$. Check: $144 + 90 + 54 + 72 = 360°$.

28 A fair number cube has 6 equally likely outcomes, making it a uniform model.

Find more at
ViewMath.com/FL-Grade7

29) $P(heads) = \frac{1}{2}$. Numbers greater than 4: $5, 6$, so $P = \frac{2}{6} = \frac{1}{3}$. $P = \frac{1}{2} \times \frac{1}{3} = \frac{1}{6}$.

30) A spinner with 4 equal sections has $\frac{1}{4} = 25\%$ chance for each section. Assigning one section as "win" gives 25%.

✔ Practice Test 8 — Answer Key

1) B	2) $37\frac{1}{2}$ cups	3) 70	4) B	5) 3%	6) C	7) A	8) C	9) B

10) $5x - 3$	11) A	12) B	13) $x \le 10$	14) $s \ge 87$; closed circle at 87, shade right

15) $1\ cm = 18\ ft$	16) Exactly one	17) A triangle	18) C	19) 10 cm	20) C	21) 6 cm

22) B	23) B	24) 1.6	25) 2 MADs	26) C	27) 144°	28) A	29) $\frac{1}{12}$	30) B

💡 Time to Learn! 💡

Review the explanations below, **especially for the questions you missed**.

Understanding why each answer is correct builds stronger problem-solving skills.

Tip: Circle any questions you got wrong, then read their explanation carefully.

📖 Practice Test 8 — Detailed Explanations

1) $k = \frac{150}{2.5} = 60$ miles per hour.

2) $k = \frac{7.5}{3} = 2.5$ cups per batch. For 15 batches: $2.5 \times 15 = 37.5 = 37\frac{1}{2}$ cups.

Find more at
ViewMath.com/FL-Grade7

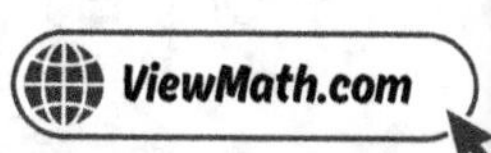

3 Divide: $56 \div 0.80 = 70$.

4 $\frac{27}{180} = \frac{p}{100}$. Cross-multiply: $180p = 2700$. Divide: $p = 15\%$.

5 $180 = 2{,}000 \times r \times 3 \boxtimes 180 = 6{,}000r \boxtimes r = 0.03 = 3\%$.

6 $\frac{|35-40|}{40} \times 100 = \frac{5}{40} \times 100 = 12.5\%$.

7 $(-25) + 13 = -12$. Then $(-12) + (-8) = -20$.

8 Difference $= 134 - (-282) = 134 + 282 = 416$ feet.

9 GCF of 24 and 16 is 8. Factor: $24x \div 8 = 3x$ and $-16 \div 8 = -2$. Result: $8(3x - 2)$.

10 Combine: $2x + 3 + 4x - 1 - x - 5$. Variable: $2x + 4x - x = 5x$. Constants: $3 - 1 - 5 = -3$. Result: $5x - 3$.

11 Two identical groups of $(x + 4)$: $2(x + 4) = 22$. Divide by 2: $x + 4 = 11$. Subtract 4: $x = 7$.

12 Multiply every term by 6: $x - 4 = 3$. Add 4: $x = 7$. Check: $\frac{7}{6} - \frac{2}{3} = \frac{7}{6} - \frac{4}{6} = \frac{3}{6} = \frac{1}{2}$ ✓.

13 Subtract 7: $\frac{x}{2} \le 5$. Multiply by 2: $x \le 10$.

14 $\frac{78+85+90+s}{4} \ge 85$. Multiply by 4: $253 + s \ge 340$. Subtract 253: $s \ge 87$. Closed circle at 87, shade right.

15 Actual length: $9 \times 6 = 54$ ft. New scale: $54 \div 3 = 18$ ft per cm.

16 AAA alone gives many, but specifying a side length fixes the size. A triangle with all 60° angles is equilateral, so all sides are 10 cm. Exactly one triangle.

Find more at
ViewMath.com/FL-Grade7

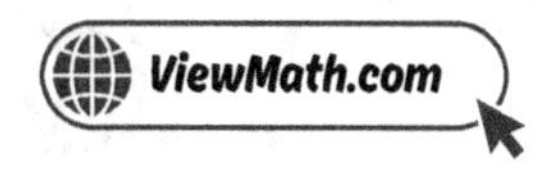

17 A vertical slice through the apex of a cone produces an isosceles triangle. The base of the triangle is a diameter of the cone's base.

18 Vertical angles are always equal. They are formed by two intersecting lines, and the opposite angles are congruent.

19 $d = C \div \pi = 62.8 \div 3.14 = 20$ cm. Then $r = 20 \div 2 = 10$ cm.

20 Two triangles: $2 \times \frac{1}{2} \times 6 \times 4 = 24$ cm^2. Three rectangles: $6 \times 8 + 5 \times 8 + 5 \times 8 = 48 + 40 + 40 = 128$ cm^2. Total: $24 + 128 = 152$ cm^2.

21 $360 = 12 \times w \times 5 = 60w$. $w = 360 \div 60 = 6$ cm.

22 The volume of a cylinder is $V = \pi r^2 h$ (area of the circular base times the height).

23 Selecting every fifth student from a list is systematic and gives all students an equal chance, making it more representative.

24 Deviations from mean: $|4 - 7| + |6 - 7| + |7 - 7| + |8 - 7| + |10 - 7| = 3 + 1 + 0 + 1 + 3 = 8$. MAD $= \frac{8}{5} = 1.6$.

25 Difference: $57 - 45 = 12$. In MADs: $\frac{12}{6} = 2$.

26 Count all the leaves: $3 + 5 + 2 = 10$ data values.

27 $\frac{24}{60} = 0.40 = 40\%$. Angle $= 0.40 \times 360° = 144°$.

28 Predicted: $\frac{5}{10} = 0.5$. Experimental: $\frac{24}{40} = 0.6$. The small difference is expected from natural variation.

29 $P(\text{tails}) = \frac{1}{2}$, $P(1) = \frac{1}{6}$. $P = \frac{1}{2} \times \frac{1}{6} = \frac{1}{12}$.

Find more at
ViewMath.com/FL-Grade7

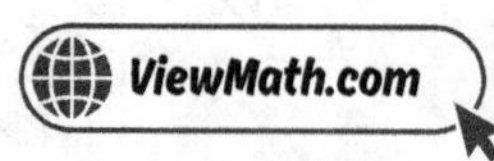

30 More trials produce more reliable estimates. The experimental probability approaches the theoretical value as trials increase.

☑ Practice Test 9 — Answer Key

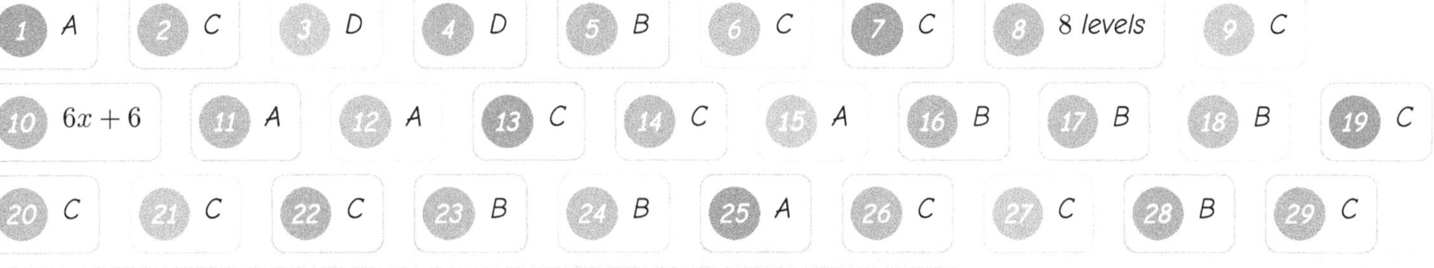

1 A **2** C **3** D **4** D **5** B **6** C **7** C **8** 8 levels **9** C

10 $6x + 6$ **11** A **12** A **13** C **14** C **15** A **16** B **17** B **18** B **19** C

20 C **21** C **22** C **23** B **24** B **25** A **26** C **27** C **28** B **29** C

30 Roll the die; let 1 and 2 represent the event (or any 2 numbers out of 6)

💡 Time to Learn! 💡

Review the explanations below, **especially for the questions you missed**.

Understanding why each answer is correct builds stronger problem-solving skills.

Tip: Circle any questions you got wrong, then read their explanation carefully.

📖 Practice Test 9 — Detailed Explanations

1 By convention, $k = \frac{y}{x} = \frac{20}{8} = 2.5$. Student B computed $\frac{x}{y}$, which is the reciprocal — not k.

2 Unit price: $\$42.50 \div 5 = \$8.50$ each. For 8: $\$8.50 \times 8 = \$68.00$.

3 $40\% = 0.40$. Multiply: $0.40 \times 150 = 60$.

Find more at
ViewMath.com/FL-Grade7

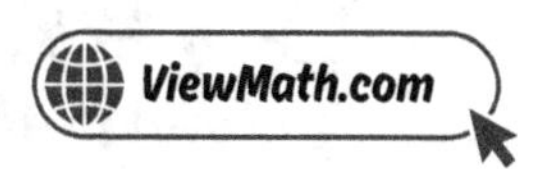

4. $\frac{60}{150} = \frac{p}{100}$. *Cross-multiply:* $150p = 6000$. *Divide:* $p = 40\%$.

5. $120 = 1{,}500 \times 0.04 \times t$ ⊠ $120 = 60t$ ⊠ $t = 2$ *years.*

6. $\frac{|8{,}000 - 7{,}500|}{7{,}500} \times 100 = \frac{500}{7{,}500} \times 100 \approx 6.67\%$.

7. *Owing money is negative. You owe* \$25 (-25) *and borrow* \$10 *more* (-10): $(-25) + (-10) = -35$.

8. $|5 - (-3)| = |5 + 3| = |8| = 8$ *levels.*

9. *GCF of 5 and 7 is 1, so* $5n + 7$ *cannot be factored further. All other expressions have a GCF greater than 1.*

10. *Add all sections:* $(3x + 5) + (2x - 3) + (x + 4) = (3x + 2x + x) + (5 - 3 + 4) = 6x + 6$.

11. *Divide by 5:* $n + 3 = -2$. *Subtract 3:* $n = -5$. *Check:* $5(-5 + 3) = 5(-2) = -10$ ✓.

12. *Subtract 3:* $0.5x = 5$. *Divide by 0.5:* $x = 10$. *Check:* $0.5(10) + 3 = 5 + 3 = 8$ ✓.

13. *Solve:* $4x > 12$, *so* $x > 3$. *Only* $x = 4$ *satisfies* $x > 3$. *Check:* $4(4) - 1 = 15 > 11$ ✓.

14. $x \leq -2$: *closed circle (includes* -2*) and shade left (values less than or equal to* -2*).*

15. *Scale ratio:* $\frac{25}{75} = \frac{1}{3}$. *New length:* $6 \times \frac{1}{3} = 2$ *cm.*

16. $7 + 10 = 17 > 15$ ✓, $7 + 15 = 22 > 10$ ✓, $10 + 15 = 25 > 7$ ✓. *SSS with valid lengths gives one unique triangle.*

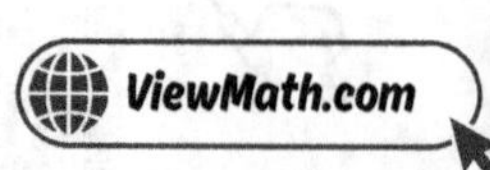

17   A cut parallel to the base of a prism always produces a shape congruent (same size and shape) to the base. For a triangular prism, this is a triangle identical to the base.

18 $2x + 10 + 3x = 180$. Combine: $5x + 10 = 180$. Subtract 10: $5x = 170$. Divide: $x = 34$.

19   A diameter passes through the center and connects two points on the circle.

20   Two bases: $2 \times 15 = 30$ cm^2. Lateral faces: $40 + 30 + 50 = 120$ cm^2. Total: $30 + 120 = 150$ cm^2.

21 $B = \frac{1}{2}(4 + 6) \times 3 = 15$ cm^2. $V = 15 \times 10 = 150$ cm^3.

22 $V = \pi r^2 h = 3.14 \times 9 \times 10 = 282.6$ cm^3.

23   Testing every light bulb would destroy them all. Sampling lets you estimate the average without testing every one.

24   Overlap means some members of each group have similar values, even if the centers differ.

25   Player A has MAD $= 3$ (vs. 7) and range $= 10$ (vs. 25), so Player A's scores are more tightly clustered around the mean.

26   With an even number of values (12), the median is the average of the 6th and 7th values.

27 $\frac{3}{24} = 0.125 = 12.5\%$.

28   Predicted: $\frac{1}{4} = 0.25$. Observed: $\frac{25}{80} = 0.3125$. They are close, which supports the model.

29   Only one outcome out of 8 is all tails (TTT). $P = \frac{1}{8}$.

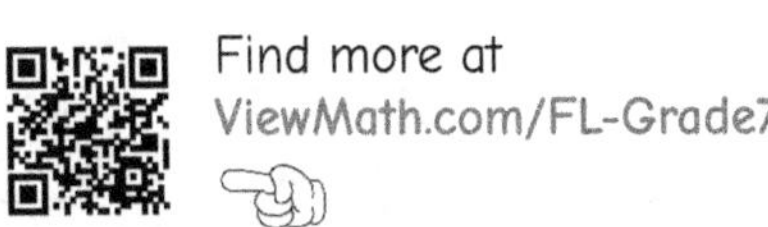

30   *Choose 2 of the 6 faces to represent the event.* $P = \frac{2}{6} = \frac{1}{3}$.

✅ Practice Test 10 — Answer Key

1   B   2   C   3   C   4   C   5   B   6   Day 1 with 25% error   7 -12

8   17 meters   9   B   10   C   11 $x = 6$ 12   A   13 $2x + 8 > 20;\ x > 6$ 14   A

15   C   16   55°   17   C   18 $a = 128°,\ b = 52°,\ c = 128°$ 19   C   20   A   21   D   22   C

23   Only listeners who choose to call participate, so it is voluntary and biased   24   A   25   B   26   34.5

27   B   28   B   29   B   30   A

💡 Time to Learn! 💡

*Review the explanations below, **especially for the questions you missed**.*

Understanding why each answer is correct builds stronger problem-solving skills.

Tip: *Circle any questions you got wrong, then read their explanation carefully.*

📖 Practice Test 10 — Detailed Explanations

1   *Use any labeled point:* $k = \frac{2}{3}$ *from* $(3, 2)$. *Check:* $\frac{4}{6} = \frac{2}{3}$ ✓.

2   *Rate:* $90 \div 2 = 45$ *mph. In 9 hours:* $45 \times 9 = 405$ *miles.*

Find more at
ViewMath.com/FL-Grade7

3 Divide the part by the whole: $18 \div 72 = 0.25 = 25\%$.

4 Cross-multiply: $100 \times n = 250 \times 16$, which gives $100n = 4000$.

5 $264 = 1{,}600 \times 0.055 \times t$ ⬚ $264 = 88t$ ⬚ $t = 3$ years.

6 Day 1: $|10 - 8|/8 = 25\%$. Day 2: $|6 - 5|/5 = 20\%$. Day 3: $|8 - 10|/10 = 20\%$. Day 1 has the largest percent error at 25%.

7 Three negatives with same sign: $4 + 4 + 4 = 12$. Keep the negative: -12.

8 $|12 - (-5)| = |12 + 5| = |17| = 17$ meters.

9 GCF of 8 and 20 is 4. Factor: $8m \div 4 = 2m$ and $20 \div 4 = 5$. Result: $4(2m + 5)$. Choice A uses 2, which is not the GCF.

10 Combine: $(5n - 2n) + (8 + 3) = 3n + 11$.

11 Total area $= 6 \times 2 \times (x + 1) = 84$. So $12(x + 1) = 84$. Divide by 12: $x + 1 = 7$. Subtract 1: $x = 6$.

12 Add 2.4: $0.6x = 3.6$. Divide by 0.6: $x = 6$. Check: $0.6(6) - 2.4 = 3.6 - 2.4 = 1.2$ ✓.

13 The left side is heavier (tips down), so $2x + 8 > 20$. Subtract 8: $2x > 12$. Divide by 2: $x > 6$.

14 Subtract 6: $-2x \leq -10$. Divide by -2 and flip: $x \geq 5$. Closed circle at 5, shade right.

15 Actual road: $7 \times 5 = 35$ km. New scale: $35 \div 3.5 = 10$. So the second map uses 1 cm $= 10$ km.

Find more at
ViewMath.com/FL-Grade7

16 $180° - 72° - 53° = 55°.$

17 Any cut parallel to the base of a prism gives a shape congruent to the base. A hexagonal prism has a hexagonal base, so the cross-section is a hexagon.

18 Vertical to 52°: $b = 52°$. Supplementary to 52°: $a = 180° - 52° = 128°$. Vertical to a: $c = 128°$.

19 $d = 2r = 2 \times \frac{3}{4} = \frac{6}{4} = 1\frac{1}{2}$ inches.

20 $SA = 6s^2$. So $s^2 = 150 \div 6 = 25$. Therefore $s = 5$ cm.

21 $V = lwh = 6 \times 4 \times 3 = 72$ cm^3.

22 $V = \pi r^2 h$. If r becomes $2r$, then $V = \pi(2r)^2 h = 4\pi r^2 h$. The volume quadruples.

23 Voluntary response samples are biased because people with strong opinions are more likely to respond.

24 Group A's histogram peaks at 6–8 hours, while Group B's peaks at 2–4 hours. Group A tends to study more.

25 Group A range $= 20$, Group B range $= 60$. Group B's data is far more spread out.

26 6 values: median $= \frac{33+36}{2} = \frac{69}{2} = 34.5.$

27 The entire circle represents the whole — 100% of the data or 360°.

28 Red covers half the spinner (180°), while Blue and Green each cover a quarter (90° each). This is a non-uniform model.

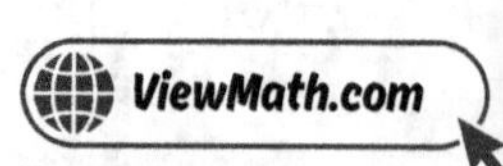

(29) *Pairs with product 12: $(2,6), (6,2), (3,4), (4,3)$ — that is 4 out of 36. $P = \frac{4}{36} = \frac{1}{9}$.*

(30) $P(\text{sum} = 7) = \frac{6}{36} = \frac{1}{6}$. *Expected: $36 \times \frac{1}{6} = 6$. The simulation gave 8, slightly above expected.*

Well done checking your answers!

Keep practicing to strengthen your skills.

Author's Final Note

I hope you enjoyed this book as much as I enjoyed writing it. Whether you are a student working through the material, a parent supporting your child's learning, or a teacher guiding your class, I have tried to make this book as clear and engaging as possible. I hope I have succeeded. If you have any suggestions for improvement, please let me know. I would love to hear from you.

The accuracy of calculations is very important to me. We have done our best, but I also expect that I have made some minor errors. Constant improvement is the name of the game. If you find any errors, please let me know. I will fix them in the next edition.

For students: Your learning journey does not end here. I have written a series of books to help you learn math. Make sure you browse through them. I especially recommend workbooks and practice tests to help you prepare for your exams.

For parents: Thank you for investing in your child's education. I encourage you to explore the companion resources available online to help support your child outside the classroom.

For teachers: Thank you for the invaluable work you do every day. I hope this book serves as a useful resource in your classroom. Feel free to reach out if you have suggestions or would like to discuss how best to use this book with your students.

I also enjoy reading your reviews. If you have a moment, please leave a review on where you found this book. It will help others find this book. If you have any questions or comments, please feel free to contact me at drNazari@ViewMath.com.

And one last thing: Remember to use online resources for additional help. I recommend using the resources on https://ViewMath.com You can find video lessons, practice problems, and more. You can also use the online companion for this book to track your progress and access additional resources.

Wishing all students the best in their studies, parents every success in supporting their children, and teachers continued inspiration in their classrooms!

Dr. A. Nazari

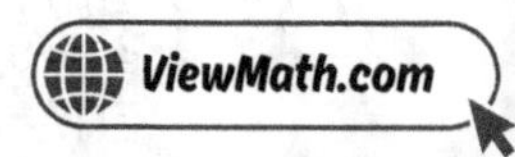

📖 *Great Job! Keep Learning with ViewMath!*

*Keep up the great work! Visit **viewmath.com/FL-Grade7** for free lessons, quizzes, and more.*

Study Guide

Workbook

Step-by-Step

3 Practice Tests

5 Practice Tests

7 Practice Tests

Find more at
ViewMath.com/FL-Grade7

www.ingramcontent.com/pod-product-compliance
Lightning Source LLC
Chambersburg PA
CBHW060158120726
48004CB00007B/1605